Lecture Notes in Control and Information Sciences

Edited by M. Thoma and A. Wyner

Vol. 62: Analysis and Optimization
of Systems
Proceedings of the Sixth International
Conference on Analysis and Optimization.
of Systems
Nice, June 19–22, 1984
Edited by A. Bensoussan, J. L. Lions
XIX, 591 pages. 1984.

Vol. 63: Analysis and Optimization
of Systems
Proceedings of the Sixth International
Conference on Analysis and Optimization
of Systems
Nice, June 19–22, 1984
Edited by A. Bensoussan, J. L. Lions
XIX, 700 pages. 1984.

Vol. 64: Arunabha Bagchi
Stackelberg Differential Games
in Economic Models
VIII, 203 pages, 1984

Vol. 65: Yaakov Yavin
Numerical Studies
in Nonlinear Filtering
VIII, 273 pages, 1985.

Vol. 66: Systems and Optimization
Proceedings of the Twente Workshop
Enschede, The Netherlands, April 16–18, 1984
Edited by A. Bagchi, H. Th. Jongen
X, 206 pages, 1985.

Vol. 67: Real Time Control of Large Scale Systems
Proceedings of the First European Workshop
University of Patras, Greece, Juli 9–12, 1984
Edited by G. Schmidt, M. Singh, A. Titli,
S. Tzafestas
XI, 650 pages, 1985.

Vol. 68: T. Kaczorek
Two-Dimensional Linear Systems
IX, 397 pages, 1985.

Vol. 69: Stochastic Differential Systems –
Filtering and Control
Proceedings of the IFIP-WG 7/1 Working Conference
Marseille-Luminy, France, March 12-17, 1984
Edited by M. Metivier, E. Pardoux
X, 310 pages, 1985.

Vol. 70: Uncertainty and Control
Proceedings of a DFVLR International Colloquium
Bonn, Germany, March, 1985
Edited by J. Ackermann
IV, 236 pages, 1985.

Vol. 71: N. Baba
New Topics in Learning Automata
Theory and Applications
VII, 231 pages, 1985.

Vol. 72: A. Isidori
Nonlinear Control Systems:
An Introduction
VI, 297 pages, 1985.

Vol. 73: J. Zarzycki
Nonlinear Prediction
Ladder-Filters for Higher-Order
Stochastic Sequences
V, 132 pages, 1985.

Vol. 74: K. Ichikawa
Control System Design based on
Exact Model Matching Techniques
VII, 129 pages, 1985.

Vol. 75: Distributed Parameter
Systems
Proceedings of the 2nd International
Conference, Vorau, Austria 1984
Edited by F. Kappel, K. Kunisch,
W. Schappacher
VIII, 460 pages, 1985.

Vol. 76: Stochastic Programming
Edited by F. Archetti, G. Di Pillo,
M. Lucertini
V, 285 pages, 1986.

Vol. 77: Detection of
Abrupt Changes in Signals
and Dynamical Systems
Edited by M. Basseville,
A. Benveniste
X, 373 pages, 1986.

Vol. 78: Stochastic
Differential Systems
Proceedings of the 3rd Bad Honnef
Conference, June 3–7, 1985
Edited by N. Christopeit, K. Helmes,
M. Kohlmann
V, 372 pages, 1986.

Vol. 79: Signal
Processing for Control
Edited by K. Godfrey, P. Jones
XVIII, 413 pages, 1986.

Vol. 80: Artificial Intelligence
and Man-Machine Systems
Edited by H. Winter
IV, 211 pages, 1986.

Lecture Notes in Control and Information Sciences

Edited by M. Thoma and A. Wyner

133

R. P. Leland

Stochastic Models for Laser Propagation in Atmospheric Turbulence

Springer-Verlag Berlin Heidelberg GmbH

Author
Robert Patton Leland
Dept. of Electrical Engineering
University of California, Los Angeles
Los Angeles, CA 90024
USA

ISBN 978-3-540-51538-8   ISBN 978-3-540-48231-4 (eBook)
DOI 10.1007/978-3-540-48231-4

Offsetprinting: Mercedes-Druck, Berlin

2161/3020-543210   Printed on acid-free paper.

Contents

1 Introduction   **1**
     1.1   A Brief Historical Perspective 1
     1.2   Objectives of This Monograph 1
     1.3   Organization 3

2 Wave Propagation In A Random Medium   **4**
     2.1   Introduction 4
     2.2   Statistical Description of Atmospheric Turbulence 4
     2.3   Classical Theory: Perturbation Methods 6
     2.4   The Parabolic Approximation 8
     2.5   Laser Beam Model 9
     2.6   The Markov Approximation 10

3 White Noise In Hilbert Spaces   **13**
     3.1   Introduction 13
     3.2   Review Of White Noise Theory 13
     3.3   Abstract Bilinear Systems 17
     3.4   Hilbert-Schmidt Operators 19
     3.5   Relation to Ito Integrals 21
     3.6   A White Noise Model For Wave Propagation 25
     3.7   An Ito Differential Equation Model 27
     3.8   The Space H^2 28
     3.9   The Space $\mathcal{F}$ 31

4 Product Formula Solutions   **33**
     4.1   Review Of Trotter-Kato Theory 33
     4.2   Convergence Of Solutions To Parabolic Equations With Weakly Convergent Coefficients 34
     4.3   Convergence Of Product Forms For Laser Propagation 37
     4.4   Product Forms As Physical Random Variables 42
     4.5   Convergence Of The Corresponding Ito Integrals 44

5 Simulation   **49**
     5.1   Simulation Problem Statement 49
     5.2   Application Of Product Formulas 50
     5.3   Generating Pseudo-Random Fields 54
     5.4   Weak Convergence Of Trigonometric Series 56

5.5 The Mutual Coherence Function 60
5.6 The Distribution Of The Irradiance Function 62
5.7 White Noise As The Limit Of An Ornstein-Uhlenbeck Process: Theory . . 70
5.8 White Noise As The Limit Of An Ornstein-Uhlenbeck Process: Simulation 76
5.9 Distortion Of The Beam 89

6 **Feynman Path Integrals**   **105**
6.1 Relation To Product Formulas 105
6.2 A Path Integral For Laser Propagation: The Feynman-Ito Equation 106
6.3 Discussion Of The Work Of K. Furutsu 108
6.4 First Order Approximate Solutions 109
6.5 Locally Linear Approximate Solutions 110
6.6 Second Order Approximate Solutions 111
6.7 Approximate First Moment 113

A **Simulation Software**   **120**
A.1 Overview of the Program PROPAPP 120
A.2 Instructions for Use 120
A.3 Software Description 121

B **Additional Simulation Results**   **127**
B.1 Fits of the Gamma Distribution for the Irradiance 127
B.2 Sample Plots Of Distorted Beams: Run 2 and Run 4 131

C **Simulation Verification**   **138**
C.1 Introduction 138
C.2 Random Number Generator 138
C.3 Finite Difference Equations 140
C.4 Product Form Approximation 141

List of Tables

5.1 Comparison of e^{A+B} and $e^A e^B$ 52

5.2 Comparison of $\frac{1}{2}(e^A e^B + e^B e^A)$ and e^{A+B} 53

5.3 Comparison of $e^{A/2} e^B e^{A/2}$ and e^{A+B} 54

5.4 Moments of the Irradiance Function 65

5.5 Moments of the Irradiance Function, Simulated and Log-Normal Prediction 66

5.6 Covariance For Discretized Ornstein-Uhlenbeck Process, $L = 5$ 77

5.7 Normalized Covariance For Ornstein-Uhlenbeck Process, $L = 1000$ . . . 77

5.8 Gaussian Parametrization of Average Irradiance $I(0, x)$ 79

5.9 Comparison Of $I(0, 0)$ Distribution Parameters For White Noise And $a = .5, .02, .01$ 85

C.1 Chi-Squared Distribution with 39 Degrees of Freedom 139

C.2 Chi-Squared Test Values 139

C.3 Sample Means 139

C.4 Sample Correlations 140

List of Figures

2.1 Approximations made for the Forward Scattering Equation 12

4.1 Typical values of $N^l(\rho)$ 38
4.2 Typical values of α^l 38

5.1 Sampling Pattern For Generating Random Turbulence Fields 55
5.2 Normalized Covariance Function Of n_1 56
5.3 Average Irradiance Function Across Boresight, $I(0, x)$, And Off Boresight, $I(.021, x)$, At 1000 m, Based On 8000 Samples. 61
5.4 Average Coherence Function Across Boresight, $\Gamma(0, 0; 0, x)$, At 1000 m, Based On 8000 Samples 63
5.5 Average Coherence Function Off Boresight,$\Gamma(0, 0; .021, x)$, At 1000 m, Based On 8000 Samples 64
5.6 Log-Normal Fit to Irradiance Distribution for $I(0, 0)$ 67
5.7 Log-Normal Fit to Irradiance Distribution for $I(.021, 0)$ 67
5.8 Log-Normal Fit to Irradiance Distribution for $I(-.021, -.021)$ 68
5.9 Gamma Fit to Irradiance Distribution for $I(0, 0)$ 69
5.10 Gamma Fit to Irradiance Distribution for $I(.021, 0)$ 69
5.11 Gamma Fit to Irradiance Distribution for $I(-.021, -.021)$ 70
5.12 Spectral Densities For Ornstein-Uhlenbeck Processes With Parameter a 71
5.13 Average Irradiance for White Noise and O-U Input $a = .5$, Across Boresight, $I(0, x)$, And Off Boresight, $I(.021, x)$, Based On 8000, 4000 And 4000 Samples, At 1000 m 78
5.14 Average Irradiance for White Noise and O-U Input $a = .01, .02$, Across Boresight, $I(0, x)$, And Off Boresight, $I(.021, x)$, Based On 8000, 4000 And 4000 Samples, At 1000 m 80
5.15 Average Coherence Function Across Boresight For White Noise And O-U Input $a = .5$ 81
5.16 Average Coherence Function Off Boresight For White Noise And O-U Input $a = .5$ 82
5.17 Average Coherence Function Across Boresight For White Noise And O-U Input $a = .01, .02$ 83
5.18 Average Coherence Function Off Boresight For White Noise And O-U Input $a = .01, .02$ 84
5.19 Cumulative Distribution Of $I(0, 0)$ For White Noise And O-U Input $a = .5$ 86

5.20 Cumulative Distribution Of $I(.021, 0)$ For White Noise And O-U Input $a = .5$ 87

5.21 Cumulative Distribution Of $I(-.021, -.021)$ For White Noise And O-U Input $a = .5$ 87

5.22 Cumulative Distribution Of $I(0, 0)$ For White Noise And O-U Input $a = .02, .01$ 88

5.23 Cumulative Distribution Of $I(.021, 0)$ For White Noise And O-U Input $a = .02, .01$ 88

5.24 Cumulative Distribution Of $I(-.021, -.021)$ For White Noise And O-U Input $a = .02, .01$ 89

5.25 Undistorted Irradiance At 250 m And 500 m, 3-D Graph 91

5.26 Undistorted Irradiance At 750 m And 1000 m, 3-D Graph 92

5.27 Undistorted Irradiance, Contour Plot, step = .05 93

5.28 Undistorted Phase At 250 m And 500 m, 3-D Graph 94

5.29 Undistorted Phase At 750 m And 1000 m, 3-D Graph 95

5.30 Undistorted Phase, Contour Plot, step = $2\pi/10$ 96

5.31 Distorted Irradiance At 250 m And 500 m, 3-D Graph, Run 1 97

5.32 Distorted Irradiance At 750 m And 1000 m, 3-D Graph, Run 1 98

5.33 Distorted Irradiance, Contour Plot, step = .05, Run 1 99

5.34 Distorted Irradiance, Contour Plot, step = .05, Run 2 100

5.35 Distorted Irradiance, Contour Plot, step = .05, Run 3 101

5.36 Distorted Phase At 250 m And 500 m, 3-D Graph, Run 1 102

5.37 Distorted Phase At 750 m And 1000 m, 3-D Graph, Run 1 103

5.38 Distorted Phase, Contour Plot, step = $2\pi/10$, Run 1 104

A.1 Module Dependency Diagram for PROPAPP 124

A.2 Module Dependency Diagram for MAKELP 125

A.3 Module Dependency Diagram for MATFLP 126

B.1 Gamma Fit of Simulated Irradiance Distribution for $I(0, .021)$ 127

B.2 Gamma Fit of Simulated Irradiance Distribution for $I(0, -.021)$ 128

B.3 Gamma Fit of Simulated Irradiance Distribution for $I(-.021, 0)$ 128

B.4 Gamma Fit of Simulated Irradiance Distribution for $I(.021, .021)$ 129

B.5 Gamma Fit of Simulated Irradiance Distribution for $I(.021, -.021)$ 129

B.6 Gamma Fit of Simulated Irradiance Distribution for $I(-.021, .021)$ 130

B.7 Distorted Irradiance At 250 m And 500 m, 3-D Graph, Run 2 131

B.8 Distorted Irradiance At 750 m And 1000 m, 3-D Graph, Run 2 132

B.9 Distorted Phase At 250 m And 500 m, 3-D Graph, Run 2 133

B.10 Distorted Phase At 750 m And 1000 m, 3-D Graph, Run 2 134

B.11 Distorted Phase, Contour Plot, step = $2\pi/10$, Run 2 135

B.12 Distorted Irradiance, Contour Plot, step = .05, Run 4 136

B.13 Distorted Phase, Contour Plot, step = $2\pi/10$, Run 4 137

C.1 Intensity and Error at 1000 m, linear input $q = .2 \times 10^{-7}$ 142

C.2 Intensity and Error at 1000 m, linear input $q = .25 \times 10^{-7}$ 144

Chapter 1

Introduction

Laser propagation in the atmosphere has been of great interest in the last two decades, especially in the areas of optical communication and tracking. Atmospheric turbulence can have a considerable effect on a laser communication or tracking system by bending the beam, causing it to miss the receiver, or by distorting the wavefront, which reduces the energy received by a heterodyning receiver. In addition, many substantial mathematical problems are raised when considering stochastic models for turbulence, and its effect on the index of refraction and wave propagation.

1.1 A Brief Historical Perspective

Wave propagation in random media has been studied for quite a long time. An excellent history of the subject is given by Strohbehn [49]. In the 1930's, scattering of sound waves by turbulence was studied by Obukhov [41] and Rytov [46]. In the 1950's interest in the twinkling of stars led to a number of studies of the effects of atmospheric turbulence. The basis for all the modern theories was developed by Tatarskii [52], [53], [51], Klyatskin [31], [32], [33] and others in the 1950's and 60's.

Despite the long history of the problem, very little has been done to analyze the models for wave propagation in a mathematically rigorous way, two exceptions being Dawson and Papanicolao [16] and Balakrishnan [7]. Part of the purpose of this monograph is to discuss, in a mathematically rigorous fashion, some of the models for wave propagation in atmospheric turbulence.

White noise theory, particularly the Hilbert space theory pioneered by Balakrishnan [3] in the early 1970's, is receiving increasing attention in problems of stochastic filtering and control. The theory diverges from that of Ito integrals when a non-linear system is considered. The problem of laser propagation in a random turbulence field provides an excellent example of a stochastic bilinear system where the differences between the Ito and white noise theories can be seen.

1.2 Objectives of This Monograph

The first major objective of this monograph is the rigorous analysis of several function space models for laser propagation in turbulence using the Hilbert space valued white noise theory of A. V. Balakrishnan [8].

Balakrishnan [7] considered solutions of the forward scattering equation,

$$\frac{\partial}{\partial t}V_t = \frac{i}{2k}\nabla^2 V_t + ikn_{1,t}V_t$$

in the framework of abstract bilinear systems of the form

$$\dot{V}_t = AV_t + B(V_t, N_t)$$

where V_t took on values in the Hilbert space $H = L_2(\mathbf{R}^2)$. Although the principal subject of this monograph is the solutions in H, solutions in the Sobolev space H^2 are also considered in order to prove smoothness properties of the solutions in H.

The Hilbert space models adequately describe the case of a propagating beam, but not the plane wave. Solutions in the Banach space of Fresnel class functions, which also includes the plane wave case $V_0 \equiv 1$, are also considered. The space of Fresnel class functions is also the appropriate context to discuss the application of Feynman path integrals.

The effect of atmospheric turbulence on a laser beam is sometimes viewed as propagation through a series of phase screens, which instantaneously distort the phase and reduce the coherence of the beam. Mathematically this phase-screen method is equivalent to the Trotter-Kato product formula. The convergence of these product formulas in a Hilbert space and Banach space context for each white noise sample path is shown.

Balakrishnan [7] considered linear functionals of the polynomials associated with the solution V and showed that some of them were physical random variables. In this monograph, These product forms are considered as functions of a Hilbert space valued white noise, and are shown to be physical random variables taking on values in the Hilbert space H. Mean square convergence of the product forms in the Hilbert space H is also demonstrated. The Ito equation for this limit is shown to be identical to the Ito equation model used by Dawson and Papanicolaou [16].

In engineering problems, white noise is often considered as the limit of a mean square continuous process as the bandwidth is expanded. The product form solutions for an Ornstein-Uhlenbeck process input are shown to be physical random variables and to converge to the product form solutions for a white noise input in mean square.

The second objective of this monograph is a digital simulation of a laser beam propagating in random turbulence. Using a product formula approximation, this simulation was performed for a beam propagating over 1000 m, in strong turbulence. The effect of expanding the bandwidth of the turbulence field along the direction of propagation to a limiting white noise was studied by means of this simulation.

The results of the simulation for the mutual coherence function are verified, and the distribution of the irradiance function is also studied. This simulation also enabled observation of the effect of turbulence on the irradiance and phase along the path of the beam.

The use of product formulas leads naturally to the relation of solutions of the forward scattering equation and Feynman path integrals. Feynman integrals are used to discuss the effect of local variations in the turbulence field on the beam intensity, to study the effects of certain assumptions on the turbulence statistics, and to obtain an approximate expression for the mean field without making the Markov, or white noise, assumption.

1.3 Organization

This monograph is divided into seven chapters. Chapter 1 is the introduction. In Chapter 2. The derivation of the physical models for wave propagation and their historical development are presented. In Chapter 3, the theory of Hilbert space valued white noise, and its application to bilinear systems and laser propagation are reviewed. The white noise model of Balakrishnan [7] and the Ito equation model of Dawson and Papanicolaou [16] are compared. In addition, alternative function spaces for modeling laser propagation are discussed. In Chapter 4. the use of product forms of the Trotter-Kato type to obtain approximate solutions to the forward scattering equation is discussed and the convergence of these solutions is shown. In Chapter 5. the convergence of functions of an Ornstein-Uhlenbeck process to functions of a white noise as the bandwidth is expanded is considered both mathematically and as a simulation study. In addition the simulated distribution of the irradiance function and several typical simulated laser beams are discussed. In Chapter 6. I discuss the use of Feynman path integrals, show their existence for the white noise model and use them to calculate various moments of the beam intensity. Chapter 7. concludes the monograph. Additional simulation results, a description of the simulation software, and verification tests for the simulation are included in the appendices.

Chapter 2

Wave Propagation In A Random Medium

2.1 Introduction

In this chapter I present a statistical model of atmospheric turbulence due to Tatarskii [52], a brief derivation of the forward scattering equation from Maxwell's equations, and a brief discussion of the classical theory. Next I present the Markov process model of Tatarskii and Klyatskin [51],[31] which introduces the notion of a white noise into the problem.

2.2 Statistical Description of Atmospheric Turbulence

Our knowledge of the statistical properties of turbulence is limited, hence we follow the turbulence model developed by Kolmogorov [30]. The turbulence field, and hence the index of refraction is assumed to be frozen in time. This is reasonable since the turbulence changes slowly compared to the frequency of light. Tatarskii [52] [53] developed a statistical model for the index of refraction variations, using Kolmogorov's turbulence model and the idea of conservative passive additives. I will present some definitions concerning random fields and then present the model for the index of refraction variations.

Random fields are often described in terms of their second order statistics, that is their mean and covariance function. If the mean and covariance are translation and rotation invariant, the random field is called homogeneous and isotropic. A slight modification of this description is used for random atmospheric turbulence.

Let X be a random field on $\mathbf{R}^n$ with zero mean.

Definition 2.2.1 *The* structure function *for a random field X is*

$$D(r_1, r_2) = E[(X(r_1) - X(r_2))^2]$$

Definition 2.2.2 *Let X be a random field on $\mathbf{R}^n$ with zero mean. X is said to be* locally homogeneous *if*

$$D(r_1, r_2) = D(r_1 - r_2)$$

for $r_1, r_2 \in G$, where G is a connected subset of $\mathbf{R}^n$.

Hence X is locally homogeneous if its structure function is translation invariant for points in G.

Definition 2.2.3 *X is said to be* locally isotropic *if, in addition,*

$$D(r_1, r_2) = D(|r_1 - r_2|)$$

for $r_1, r_2 \in G$, where G is a connected subset of $\mathbf{R}^n$.

Hence X is locally isotropic if its structure function is rotation invariant for points in G.

Usually we look at the covariance function of a random field and talk about it being isotropic and homogeneous on all of $\mathbf{R}^n$. In the case of random turbulence using the structure function minimizes the effect of 'low frequency' variations which are not of interest to us. The 'local' part of the definitions is necessary because in nature the turbulence field is not statistically stationary over larger distances, particularly changes in altitude. However locally the assumption of an isotropic homogeneous random field is a good model for turbulence. In the analysis that follows the random field will be assumed to be homogeneous, and not just locally so, with the understanding that we are primarily interested in a bounded region.

Tatarskii's model for the index of refraction n assumes that

$$n = 1 + n_1$$

where n_1 is a zero mean locally isotropic homogeneous random field on $\mathbf{R}^3$ with structure function:

$$D(r) = C_n^2 r^{2/3}, l_0 < r < L_0 \tag{2.1}$$

l_0 and L_0 are called the inner and outer scales and describe the maximum and minimum size of the eddies that are taken into account by the model. The value of $D(r)$ outside this region is simply not known.

Let R be the covariance function for n_1 and Φ_n the corresponding spectral density. then R and Φ_n satisfy:

$$
\begin{aligned}
R(r) &= \int_{\mathbf{R}^3} e^{i\lambda \cdot r} \Phi_n(\lambda) d\lambda \\
\Phi_n(\lambda) &= \frac{1}{(2\pi)^3} \int_{\mathbf{R}^3} e^{-i\lambda \cdot r} R(r) \, dr
\end{aligned}
$$

when the integrals exist. The structure function and the spectral density are related by

$$
\begin{aligned}
D(r) &= \int_{\mathbf{R}^3} 2(1 - \cos(\lambda \cdot r)) \Phi_n(\lambda) \, d\lambda \\
\Phi_n(\lambda) &= \frac{1}{16\pi^2 |\lambda|^2} \int_{\mathbf{R}^3} \sin(\lambda \cdot r) \lambda \cdot \nabla D(r) \, dr
\end{aligned}
$$

Under the Kolmogorov turbulence model, the corresponding spectral density for n_1 is given by:

$$\Phi_n(\lambda) = .033 C_n^2 |\lambda|^{-11/3}, 2\pi/L_0 < |\lambda| < 2\pi/l_0$$

Since it is not known what P should be outside this range of frequencies, various extensions have been used. It should be pointed out that there is no data that would indicate a preference between any of these extensions. The Von Karman spectrum,

$$\Phi_n(\lambda) = .033 C_n^2 \frac{1}{(\lambda_0^2 + \lambda^2)^{11/6}} \tag{2.2}$$

accounts for an attenuation of the lower frequencies. Tatarskii used a modified Von Karman spectrum, which also attenuates the higher frequencies.

$$\Phi_n(\lambda) = .033 C_n^2 \frac{e^{\frac{\lambda^2}{\lambda_m^2}}}{(\lambda_0^2 + \lambda^2)^{11/6}} \tag{2.3}$$

Unless stated otherwise the analysis below use the model developed by Tatarskii with the modified Von Karman spectrum.

2.3   Classical Theory: Perturbation Methods

In the physics literature, discussion of wave propagation begins with Maxwell's equations in differential form:

$$\nabla \times E = -\frac{\partial}{\partial t}\mu H$$
$$\nabla \times H = \frac{\partial}{\partial t}\epsilon E$$
$$\nabla \cdot (\epsilon E) = 0$$
$$\nabla \cdot H = 0$$

where:
$E(x,y,z;t) \in \mathbf{R}^3$ is the electric field.
$H(x,y,z;t) \in \mathbf{R}^3$ is the magnetic field.
μ is the permitivity, assumed to be a constant.
$\epsilon(x,y,z)$ is the dielectric constant, assumed to be constant in time.
It is assumed that the laser is monochromatic, that is it operates at a single frequency, hence the time dependence of E and H is assumed to be sinusoidal and is described by:

$$E(x,y,z;t) = Re(E(x,y,z)e^{i\omega t})$$
$$H(x,y,z;t) = Re(H(x,y,z)e^{i\omega t})$$

where $E(x,y,z), H(x,y,z) \in C^3$.
Hence equations 2.4 and 2.4 become:

$$\nabla \times E = -i\omega\mu H \tag{2.4}$$

$$\nabla \times H = i\omega\epsilon E \tag{2.5}$$

It is possible to obtain a single equation for E, since:

$$\nabla \times \nabla \times E = \nabla \times (-i\omega\mu H) = \omega^2\mu\epsilon E = k^2n^2E \tag{2.6}$$

where k is the wave number

$$k = \frac{2\pi}{\lambda}$$

λ being the wavelength. n is the index of refraction which will be the random term. Also:

$$\nabla \times \nabla \times E = -\nabla^2E + \nabla(\nabla \cdot E) \tag{2.7}$$

From Equation 2.4 we have:

$$0 = \nabla \cdot (\epsilon E) = E \cdot \nabla\epsilon + \epsilon(\nabla \cdot E) \tag{2.8}$$

Hence,

$$\nabla \cdot E = -E \cdot \nabla \log \epsilon \tag{2.9}$$

From equations 2.6, 2.7 and 2.9 we have:

$$\nabla^2E + k^2n^2E + \nabla(E \cdot \nabla \log \epsilon) = 0 \tag{2.10}$$

The third term in Equation 2.10, $\nabla(E \cdot \nabla \log \epsilon)$, is called the depolarization term. Tatarskii [50] has shown that this term is negligible and can safely be ignored. Without it, Equation 2.10 can be written as three scalar partial differential equations that can be considered separately. Henceforward E is taken to be complex scalar valued in the equation:

$$\nabla^2E + k^2n^2E = 0 \tag{2.11}$$

A fine review of the classical theory of wave propagation in a random medium is found in Clifford [13]. In the classical theory it is assumed that $n^2 - 1 \approx 2n_1$ hence Equation 2.11 can be written as:

$$\nabla^2E + k^2E + 2kn_1E = 0 \tag{2.12}$$

The method of small perturbations assumes that E can be written as:

$$E = E_0 + E_1 + \ldots \tag{2.13}$$

where E_j is the jth order perturbation in n_1. The first two terms in this series satisfy:

$$\nabla^2E_0 + k^2E_0 = 0 \tag{2.14}$$

$$\nabla^2E_1 + k^2E_1 + 2k^2n_1E_0 = 0 \tag{2.15}$$

Convergence of the perturbation series is unknown, however usually only the first two terms are kept. This implies that E will be Gaussian if n_1 is Gaussian. The method of small perturbations gives adequate results provided the turbulence is sufficiently weak and the propagation distance is not greater than 100m. These limitations severely constrain the applicability of the method of small perturbations to communication problems.

2.4 The Parabolic Approximation

Equation 2.12 is difficult to work with so it is generally approximated by a parabolic equation. This is known as the *parabolic approximation*. Denote:

$$V(x, y, z) = E(x, y, z)e^{ikz} \qquad (2.16)$$

Then from Equation 2.12, V must satisfy:

$$\frac{\partial^2}{\partial z^2}V + 2ik\frac{\partial}{\partial z}V + \nabla^2 V + 2k^2 n_1 V = 0 \qquad (2.17)$$

where $\nabla^2 = \frac{\partial^2}{\partial x^2} + \frac{\partial^2}{\partial y^2}$. V is assumed to change slowly compared to e^{ikz} hence

$$\mid \frac{\partial^2}{\partial z^2}V \mid \ll k \mid \frac{\partial}{\partial z}V \mid \qquad (2.18)$$

The parabolic approximation consists in dropping the second order term $\frac{\partial^2}{\partial z^2}V$ in Equation 2.17 to obtain a parabolic equation.

$$\frac{\partial}{\partial z}V = \frac{i}{2k}\nabla^2 V + ikn_1 V \qquad (2.19)$$

This equation is known as the forward scattering equation or the parabolic equation and is the basis for the rest of the analysis in this monograph. Justification of the parabolic approximation can be found in Tatarskii [53].

The parabolic approximation is based mainly on the physical assumption that the scattering angles are small, and there is no backscatter. The solution to Equation 2.17 can be expressed in integral form as:

$$V(r) = \frac{1}{4\pi} \int \frac{e^{ik|r-r'|}}{|r - r'|} e^{ik(z'-z)} 2k^2 n_1(r') V(r') \, dr' \qquad (2.20)$$

Now if $r = (x, y, z)$, take $\rho = (x, y)$. Then

$$
\begin{aligned}
|r - r'| &= |z - z'|\sqrt{1 + \frac{|\rho - \rho'|^2}{|z - z'|^2}} \\
&\approx |z - z'|(1 + \frac{|\rho - \rho'|^2}{2|z - z'|^2}) \\
&= |z - z'| + \frac{|\rho - \rho'|^2}{2|z - z'|}
\end{aligned}
$$

The no backscatter assumption implies that we always take $z \geq z'$, and the small scattering angle assumption implies that $|z - z'| \gg |\rho - \rho'|$, hence the approximation above is valid. We make the approximation

$$\frac{e^{ik|r-r'|}}{|r - r'|} \approx \frac{e^{ik(z-z')+ik|\rho-\rho'|^2/2(z-z')}}{z - z'}$$

The integral equation for V is now

$$V(r) = \int_0^z \int_{\mathbf{R}^2} \frac{-2ik}{4\pi(z-z')} e^{ik\frac{|\rho-\rho'|^2}{z-z'}} (ik\ n_{1,z'}(\rho'))V_{z'}(\rho')\ dz\ d\rho \qquad (2.21)$$

which is the solution to the forward scattering equation 2.19. Hence the Greens function for the forward scattering equation approximates the original.

It should be noted that the parabolic approximation does not give an approximate solution for E but rather represents a different model for E which yields moments that approximate the moments of E. Equation 2.11 does not yield a solution which depends continuously on initial conditions in any normed linear space, whereas the Equation 2.19 represents a well posed Cauchy problem in several Hilbert and Banach spaces. Justification of this approximation and others are usually made in terms of the moments and not in terms of the solutions to the equations.

An alternative perturbation method was used by Rytov [46] in his investigation of the diffraction of light by acoustic waves. Rytov took $V = e^\psi$ and obtained the equation:

$$2ik\frac{\partial \psi}{\partial z} = \nabla^2\psi + 2\nabla\psi \cdot \nabla\psi + 2k^2 n_1 \qquad (2.22)$$

The method of smooth perturbations, or Rytov's method, consists of taking a perturbation series for ψ in terms of n_1. Although this method appeared to be more promising than the method of small perturbations, it turned out to have the same limitations, that is only in weak turbulence and only for distances under about 100m.

In either of the perturbation methods, using the first two terms corresponds to the physical assumption that any ray of light is scattered only once. In stronger turbulence or over longer distances, this assumption breaks down and multiple scattering becomes more important. Because of these limitations, other techniques that accounted for multiple scattering were developed. All of these techniques make the same assumptions and lead to the same results for moments of V. In this monograph I consider only the Markov approximation, following the work of Tatarskii and Klyatskin.

2.5   Laser Beam Model

In the framework of the forward scattering equation 2.19, it is possible to define a model for a laser beam. A standard model is to assume that the beam is Gaussian in shape. This is not a probabilistic description of the beam, but rather a statement about the initial condition $V(x, y, 0)$ of Equation 2.19. The following description of the Gaussian beam with a flat wavefront is taken from Ishimaru [24].

Let

$$\rho = (x, y)$$

then

$$V(0, \rho) = \exp(-|\rho|^2/\alpha)$$

With this initial condition, the solution to Equation 2.19 when $n_1 \equiv 0$ is

$$V(z, \rho) = \frac{\alpha}{\alpha + i2z/k} \exp\{-|\rho|^2/(\alpha + i2z/k)\}$$

According to Ishimaru this model is valid for $z \ll 8k^3\alpha^2$ For a beam with an initial width of $\alpha^{\frac{1}{2}} = .01m$ this is about $z \ll 8 \times 10^{17}m$, which holds for any laser propagation problem we are likely to consider. This is the beam model used for the simulations in this monograph.

An alternative model that is often considered is the plane wave $V(0, \rho) = 1$. The beam width is assumed to be much larger than the wavelength, hence the plane wave approximates the beam in a small area. The plane wave case is considered briefly in Chapters 3 and 4. It is difficult to simulate due to the problem of choosing appropriate boundary conditions.

2.6   The Markov Approximation

The limitations of the perturbation methods led to attempts to calculate moments of V by other means. If n_1 is modelled as an isotropic random field it is impossible to calculate the moments exactly. Tatarskii and Klyatskin assumed that the index of refraction deviations n_1 were uncorrelated in the direction of propagation, hence:

$$E(n_1(z_1, \rho_1)n_1(z_2, \rho_1)) = \delta(z_1 - z_2)A(\rho_1 - \rho_2) \qquad (2.23)$$

A is related to our original covariance model for n_1 by:

$$A(\rho) = \int_{-\infty}^{\infty} R(\rho, z) \, dz \qquad (2.24)$$

and the spectral density corresponding to $\mathbf{A}$ is

$$\Phi(\lambda) = \Phi_n(\lambda, 0), \lambda \in \mathbf{R}^2 \qquad (2.25)$$

To illustrate the advantages of the Markov approximation we calculate the first moment of V. Equation 2.19 has solution satisfying

$$V(z) = V(0) + \int_0^z \frac{i}{2k}\nabla^2 V(t)dt + ik\int_0^z n_1(t)V(t)dt$$

For any random variable denote:

$$E(x) = <x>$$

then

$$<V_z> = V_0 + \int_0^z \frac{i}{2k}\nabla^2 <V_t> dt + \int_0^z ik <n_{1t}V_t> dt$$

The Furutsu-Novikov formula (Novikov [40]) states that for functionals R of a gaussian process X,

$$<X(r)R(X)> = \int_{\mathbf{R}^n} <X(r)X(r')><\frac{\delta R(X)}{\delta X(r')}> dr' \qquad (2.26)$$

where $\frac{\delta R}{\delta X}$ denotes the variational derivative. In our case this is

$$<V_z(\rho)n_{1z}(\rho)> = \int_0^z \int_{\mathbf{R}^2} <n_{1z}(\rho)n_{1s}(\rho')><\frac{\delta V_z(\rho)}{\delta n_{1s}(\rho')}> ds \, d\rho'$$

The variational derivative satisfies:

$$\frac{\delta V_t(\rho)}{\delta n_{1s}(\rho')} = \int_s^t \frac{i}{2k} \nabla^2 \frac{\delta V_u(\rho)}{\delta n_{1s}(\rho')} \, du + ik \int_s^t \frac{\delta V_u(\rho)}{\delta n_{1s}(\rho')} n_{1u}(\rho) \, du + ik V_t(\rho) \delta(\rho - \rho')$$

Hence,

$$\frac{\delta V_t(\rho)}{\delta n_{1t}(\rho')} = ik V_t(\rho) \delta(\rho - \rho')$$

which is all we need because of the delta function in the covariance of n_1. Hence,

$$< V_t n_{1t} > = \frac{1}{2} ik A(0) < V_t >$$

therefore we obtain the following equation for $< V >$:

$$\frac{\partial}{\partial z} < V_z > = \frac{i}{2k} \nabla^2 < V_z > - \frac{1}{2} k^2 A(0) < V_z > \qquad (2.27)$$

The fact that only $\frac{\delta V_t}{\delta n_{1t}}$ was needed is a direct consequence of the Markov approximation.

The Markov approximation yields results for the moments that are equal to those obtained by solving for the exact moment equations and finding approximate solutions to these, so it can be justified on this basis. Tatarskii and Klyatskin [33] also found a sequence of equations for the moments which can be closed off by applying the Markov approximation in the nth equation. They then used this to show that the Markov approximation gives good results for the first two moments when the propagation distance is much larger than the outer scale L_0.

The approximations made to get to this point are illustrated in Figure 2.6. It should be noted that the parabolic and Markov approximations are made in that order, and cannot be reversed in any reasonable way.

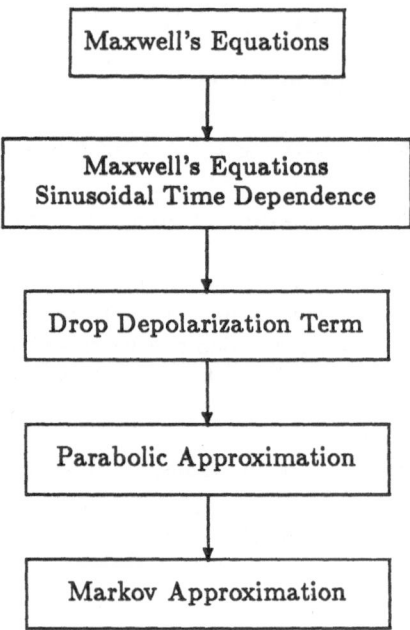

Figure 2.1: Approximations made for the Forward Scattering Equation

Chapter 3

White Noise In Hilbert Spaces

3.1   Introduction

The theory of white noise processes with sample paths in a Hilbert space is somewhat nonstandard but has been gaining greater acceptance and interest in recent years. Models using white noise tend to retain sample path properties, such as absolute continuity, that are true of the physical process being modeled. This permits a straightforward implementation of stochastic filtering and control algorithms that is not possible with Ito integrals. White noise models have been used by Balakrishnan [3,8] and Kallianpur and Karandikar [27] for problems of stochastic filtering and control. An extensive discussion of white noise theory can be found in Balakrishnan [8]. In this chapter I include a brief summary of white noise theory and discuss the previous work on its application to the laser propagation problem, as well as the Ito equation model of Dawson and Papanicolaou [16]. The white noise and Ito approaches are compared, and white noise solutions to the forward scattering equation are considered in alternative function spaces.

3.2   Review Of White Noise Theory

The theory of random variables begins with a probability triple $[\Omega, \beta, P]$ where Ω is a sample space β is a sigma algebra of sets in Ω and P is a probability measure defined on β. The random variables are the functions on Ω that are measureable with respect to β. If $\Omega = H$ where H is an infinite dimensional separable Hilbert space, and β is composed of the Borel sets of H, it is not possible to define such a triple for Gaussian white noise. Balakrishnan [8] developed a theory for such *weak random variables* and functionals of them in terms of measures on the cylinder sets of H.

Fundamental to this theory is the idea that white noise N is considered as a limit of well defined random variables N_k. A function of white noise, $f(N)$, is considered as the limit of the sequence $f(N_k)$. If the limit of the characteristic functionals for $f(N_k)$ is the characteristic functional for a random variable, $f(N)$ can be considered as such. In engineering applications, this corresponds to considering white noise to be the limit of a bandlimited stochastic process as the bandwidth is expanded to infinity.

Let H be a separable Hilbert space. Some of the definitions and theorems that are central to white noise theory are presented below. In general, the proofs are omitted, but can be found in [8].

Definition 3.2.1 *A set O is a* cylinder set *in H if O is of the form*

$$O = \{y \in H \, | \, [(y, e_1), \cdots (y, e_n)] \in B\}$$

where $n < \infty$ and B is a Borel set in $\mathbf{R}^n$. The cylinder sets form an algebra of subsets of H which is denoted by C.

Definition 3.2.2 *A* cylinder measure *μ on the cylinder sets of H is defined for all sets in C. If H_n is an n-dimensional subspace of H, $n < \infty$ and O is a cylinder set with base in H_n then there exists a countably additive measure ν_n on the Borel sets of H_n such that:*

$$\mu(O) = \nu_n(base[O])$$

for all such O . Note that a cylinder measure need not be countably additive on the Borel sets of the infinite dimensional Hilbert space H.

Definition 3.2.3 *A cylinder measure μ can be extended to be* countably additive on H *if there is a countably additive measure ν on the Borel sets of H such that $\mu(O) = \nu(O)$ for all cylinder sets O.*

A necessary and sufficient condition for when a cylinder measure can be extended to be countably additive is given in the following theorem.

Theorem 3.2.4 *A cylinder measure μ on a separable Hilbert space H can be extended to be countably additive on the Borel sets of H if and only if for every cylinder set A, if $A \subset \cup_{n=1}^{\infty} A_n$ where A_n are cylinder sets, then*

$$\mu(A) \leq \sum_{n=1}^{\infty} \mu(A_n)$$

Theorem 3.2.4 can be used to prove the following characterization of cylinder measures that can be extended to be countably additive.

Theorem 3.2.5 *A cylinder measure μ on a separable Hilbert space H can be extended to be countably additive on H if and only if for each $\epsilon > 0$ there exists a closed bounded set K such that the outer measure of K, $\mu_e(K) > 1 - \epsilon$.*

Since the characteristic functional involves only a one - dimensional projection of elements of the Hilbert space H we can define characteristic functionals with respect to cylinder measures.

$$C_\mu(h) = \int_H e^{i[h,x]} \, d\mu(x)$$

The theorem of Sazanov, which is stated here without proof, gives necessary and sufficient conditions on the characteristic functional for extending a cylinder measure to be countably additive.

Definition 3.2.6 *Let H be a real separable Hilbert space. The S topology is the weakest topology on H containing all the sets of the form*

$$\{h|[R(h-g),(h-g)] < \epsilon\}$$

where R ranges over all the nuclear operators on H.

Theorem 3.2.7 (Sazanov) *Let H be a real separable Hilbert space and μ a cylinder measure on H with characteristic functional C. μ can be extended to be countably additive on the Borel sets of H if and only if C is continuous in the S topology.*

Define the covariance operator of an H valued random variable η to be R such that

$$E([x,\eta][\eta,y]) = [Rx,y]$$

Then the characteristic function for a zero-mean Gaussian H - valued random variable η has the form:

$$C(h) = e^{-\frac{1}{2}[Rh,h]}$$

where R is the covariance of η. In the special case of Gaussian measures a more immediate condition for extension of cylinder measures exists.

Definition 3.2.8 *An operator R on a separable Hilbert space H is nuclear if, for all complete orthonormal sequences $\{e_k\}$ and $\{f_k\}$:*

$$\sum_{k=1}^{\infty} |[Re_k, f_k]| < \infty$$

Theorem 3.2.9 *Let μ be a Gaussian cylinder measure on a real separable Hilbert space H with covariance operator R. μ can be extended to be countably additive on H if and only if R is nuclear.*

It is trivial to argue that $C(h)$ is continous in the S topology if and only if R is nuclear.

Before discussing Hilbert space valued white noise it is necessary to introduce the notion of a weak random variable. Denote by $[H, \mathcal{C}, \mu]$ the "probability triple" where H is a real separable Hilbert space, $\mathcal{C}$ is the collection of cylinder sets in H and μ is a cylinder measure on H.

Definition 3.2.10 *Let H and H_r be real separable Hilbert spaces. A function $f : H \to H_r$, is a weak random variable if for each $h \in H_r$ the cylinder measure μ is defined and countably additive on sets of the form:*

$$\{\omega|[f(\omega), h] \in B\}$$

where B is any Borel set in $\mathbf{R}$.

Definition 3.2.11 *Let H be a separable Hilbert space. Gaussian white noise on $[H, \mathcal{C}, \mu]$ is the weak random variable $f(\omega) = \omega$, where μ is the standard Gaussian measure on H with characteristic functional $C_\mu(h) = e^{-\frac{1}{2}\|h\|^2}$.*

Note that white noise can be viewed as an H valued Gaussian random vector with the identity operator as its covariance. Note that this cannot correspond to a random variable with a countably additive measure because the identity is not nuclear in infinite dimensional spaces.

We are primarily interested in functions of white noise. Let f be a continuous mapping from H into a complete metric space. Let P_n be a projection operator on H with finite dimensional range. Then the function $f_n(\omega) = f(P_n\omega)$ is clearly a random variable with a well defined probability measure since for any Borel set O, $f_n^{-1}(O)$ is a cylinder set in H. These ideas are defined more formally below.

Definition 3.2.12 *Let H be a separable Hilbert space and X a Banach space and let f be a mapping of H into X. Let $\{P_n\}$ be a sequence of finite dimensional projections on H converging strongly to the identity. If the tame functions:*

$$f(P_n\omega)$$

are Cauchy in probability for all such sequences $\{P_n\}$ then f is called a physical random variable.

Physical random variables are significant in several ways. First, define

$$C(h) = \lim_{n\to\infty} \int_H e^{i<f(P_n\omega),h>} d\mu$$

where $< \cdot , \cdot >$ is the appropriate bilinear form. If f is a physical random variable then $C(h)$ is the characteristic function of a countably additive measure on X. Physical random variables are also significant as limits of functionals of finite dimensional random vectors. It is reasonable to try to simulate them on a digital computer.

If H is the Hilbert space $L_2[0,T]$, a simple example of an H valued physical random variable would be X where

$$X_t = \int_0^t N_s \, ds$$

Note that X has absolutely continuous sample paths. The corresponding Ito integral has the Wiener process W_t as a solution, which is continuous but not differentiable. An example of a nonlinear function that is a physical random variable is $[LN, N]$, where L maps H into H and $L + L^*$ is nuclear, which is true when the eigenvalues of $L + L^*$ are summable. If

$$(LN)_t = \int_0^t N_s \, ds$$

then $L + L^*$ is nuclear, hence

$$[LN, N] = \int_0^T \int_0^t N_t N_s \, dsdt$$

is a physical random variable with expected value

$$E[LN, N] = \frac{1}{2} Tr.(L + L^*) = \frac{1}{2}T$$

Note that for the Ito polynomial $I_2(L)$,

$$E[I_2(L)] = E \int_0^T \int_0^t dW_s dW_t = 0$$

hence the Ito and white noise formulations are not the same in the non-linear case.

3.3 Abstract Bilinear Systems

Solutions of abstract bilinear systems with 'multiplicative' white noise input were studied by Balakrishnan [4]. The definitions and results of [4] are presented in part in this section. The forward scattering equation 2.19 is considered here as an abstract bilinear system. To establish some notation and concepts I first discuss the abstract bilinear system and develop the wave propagation model in a later section.

Let H and H_r be separable Hilbert spaces. Consider the bilinear equation:

$$\dot{V}_t = AV_t + B(V_t, N_t) \ , \ V_0 \ given \tag{3.1}$$

where:

$$V \ \in \ L_2([0,T];H) \tag{3.2}$$
$$N \ \in \ L_2([0,T];H_r) \tag{3.3}$$

A is a closed linear operator on H with dense domain that generates a C_0 semigroup S_t such that:

$$\|S_t X\| \le M\|X\| \ , \ \forall t \in [0,T], \ X \in H \tag{3.4}$$

B is a bilinear form on $H \times H_r$ such that

$$\|B(X,N)\| \le \|B\| \cdot \|X\| \cdot \|N\| \tag{3.5}$$

I will also use the notation:

$$B(X,N) = L(N)X = \mathcal{L}_X N$$

where L is a continuous linear map from H_r into the bounded linear operators on H and $\mathcal{L}$ is a continuous linear map from H into the bounded linear operators mapping H_r into H.

In general, unless A is bounded, Equation 3.1 will not be satisfied for all initial conditions V_0. Therefore the mild solution is considered:

$$V_t = S_t V_0 + \int_0^t S_{t-\tau} B(V_\tau, N_\tau) \ d\tau \tag{3.6}$$

or, equivalently:

$$V_t = S_t V_0 + \int_0^t S_{t-\tau} L(N_\tau) V_\tau \ d\tau \tag{3.7}$$

Define $V^0 \in \mathcal{H}$ as

$$V_t^0 = S_t V_0$$

and the operator $K(N) : \mathcal{H} \to \mathcal{H}$ as

$$g \ = \ K(N)f \ , \ f,g \in \mathcal{H} \tag{3.8}$$
$$g_t \ = \ \int_0^t S_{t-s} L(N_s) f_s \ ds \tag{3.9}$$

Equation 3.7 can be expressed as an equation in H_s:

$$V = V^0 + K(N)V \tag{3.10}$$

Theorem 3.3.1 *If A and B satisfy the conditions in Equations 3.4 and 3.5 on the separable Hilbert space H, then Equation 3.10 has a unique solution in $\mathcal{H}$ for all $V^0 \in \mathcal{H}$. In addition, both the solution V and its values at specific times V_t are continuous functions of the input N.*

Proof
Equation 3.10 can be rewritten as

$$[I - K(N)]V = V^0$$

If the series:

$$\sum_{n=0}^{\infty} K^n(N)X \qquad (3.11)$$

converges strongly for all $X \in \mathcal{H}$ then $[I - K(N)]^{-1}$ exists and a solution to Equation 3.10 can be expressed as:

$$V = [I - K(N)]^{-1}V^0 = \sum_{n=0}^{\infty} K^n(N)V^0 \qquad (3.12)$$

For each $n > 0$, $K^n(N)$ can be expressed:

$$(K^n(N)X)_t =$$
$$\int_0^t \int_0^{s_1} \cdots \int_0^{s_{n-1}} S(t - s_1)L(N(s_1))S(s_1 - s_2) \cdots$$
$$\cdots S(s_{n-1} - s_n)L(N(s_n))X(s_n) \, ds \qquad (3.13)$$

and

$$\|(K^n(N)X)_t\| \leq \int_0^t \int_0^{s_1} \cdots \int_0^{s_{n-1}} M^n \|B\|^n \|N(s_1)\| \cdots \|N(s_n)\| \|X(s_n)\| \, ds$$
$$\leq (M\|B\|)^n \frac{t^{n-\frac{1}{2}}}{(n-1)!} \|X\|$$

Hence,

$$\|K^n(N)X\| \leq (M\|B\|\|N\|T)^n \frac{1}{(n-1)!\sqrt{2n}} \|X\|$$

Hence the series in Equation 3.12 converges because of the $\frac{1}{(n-1)!}$ factor. Note that the series converges in H for each t and in $\mathcal{H}$. Hence $[I - K(N)]^{-1}$ exists and is bounded by

$$\|[I - K(N)]^{-1}\| \leq f(\|N\|)$$

and

$$\|([I - K(N)]^{-1}X)_t\| \leq g(\|N\|)\|X\|$$

where f and g are continuous and $f(0) = g(0) = 0$. Hence

$$V = [I - K(N)]^{-1}V^0$$

is the unique solution to Equation 3.7.

Let $V(N)$ denote the solution for noise input N. Then

$$
\begin{aligned}
V(N_1) - V(N_2) &= [I - K(N_1)]^{-1}V^0 - [I - K(N_2)]^{-1}V^0 \\
&= [I - K(N_1)]^{-1}K(N_1 - N_2)[I - K(N_2)]^{-1}V^0
\end{aligned}
$$

Hence

$$
\|V(N_1) - V(N_2)\| \leq f(\|N_1\|)M\|B\|T \ \|N_1 - N_2\| \ f(\|N_2\|)\|V^0\|
$$

Hence the solution V is continuous in N. In addition for each t,

$$
\|V_t(N_1) - V_t(N_2)\| \leq g(\|N_1\|)M\|B\|T \ \|N_1 - N_2\| \ f(\|N_2\|)\|V^0\|
$$

Hence for each t, $V_t(N)$ is continuous in N. $\square$

3.4 Hilbert-Schmidt Operators

In the present context Hilbert-Schmidt operators are important because if L is H.S. and N is a white noise then LN is a physical random variable with a nuclear (trace class) covariance LL^*. Also, if L is a Hilbert-Schmidt map of $\otimes^n \mathcal{W}$ into a separable Hilbert space H_1,

$$
L(N^1, \cdots, N^n) = \int \cdots \int L(t_1, \cdots, t_n; N^1(t_1), \cdots, N^n(t_n)) \ dt
$$

then for the Ito polynomial

$$
I_n(L) = \int \cdots \int L(t_1, \cdots, t_n; dW(t_1), \cdots, dW(t_n))
$$

we have,

$$
E[\|I_n(L)\|^2] = \|L\|_{HS}^2
$$

These terms are defined more precisely below.

Definition 3.4.1 *A bounded linear operator L from a separable Hilbert space H_1 into a possibly different Hilbert space H_2 is said to be Hilbert-Schmidt if for any complete orthonormal sequence $\{e_k\}$ in H_1 the following property holds:*

$$
\sum_{k=1}^{\infty} \|Le_k\|^2 < \infty
$$

Proposition 3.4.2 *Let H_1 be a separable Hilbert space, H_2 be a Hilbert space and L be a Hilbert -Schmidt operator from H_1 into H_2. The series summation:*

$$
\sum_{k=1}^{\infty} \|Le_k\|^2 < \infty
$$

is the same for any complete orthonormal sequence $\{e_k\}$ in H_1.

Definition 3.4.3 *Let H_1 be a separable Hilbert space, H_2 be a Hilbert space and $\{e_k\}$ be a CON sequence in H_1. Let L, M be Hilbert -Schmidt operators from H_1 into H_2. Define the inner product of L and M to be*

$$\sum_{k=1}^{\infty} [Le_k, Me_k]$$

and the Hilbert-Schmidt norm of L to be

$$\|L\|_{HS}^2 = \sum_{k=1}^{\infty} \|Le_k\|^2$$

It can be shown that the Hilbert-Schmidt operators with this inner product form a Hilbert space.

The operators $K^n(N)X$ can be interpreted as polynomial maps from $\mathcal{W}$ to $\mathcal{H}$ or they can be viewed as linear maps from the tensor product space $\mathcal{W} \otimes \mathcal{W} \otimes \cdots \otimes \mathcal{W}$ into $\mathcal{H}$. The Hilbert-Schmidt property of the operators $K^n(\cdot)X$ defined in the previous section was shown by Balakrishnan [4].

Theorem 3.4.4 *Let H and H_r be separable Hilbert spaces and let*

$$\begin{aligned} \mathcal{H} &= L_2[(0,T); H] \\ \mathcal{W} &= L_2[(0,T); H_r] \end{aligned}$$

be Hilbert spaces with the usual inner product. Let S_t be a semigroup on H such that $\|S_t\| \leq M$, $\forall t \in (0,T)$. Let B be a bilinear operator mapping $H \times H_r$ into H such that $\|B(Y, \cdot)\|_{HS} \leq C\|Y\|$ for any Y in H, and let X denote any element of $\mathcal{H}$. Then if $K : \mathcal{W} \times \mathcal{H} \rightarrow \mathcal{H}$ is defined as

$$[K(N)X]_t = \int_0^t S_{t-s} B(X_s, N_s) \, ds$$

The mappings $K^n(\cdot)X$ from $\otimes^n \mathcal{W}$ into $\mathcal{H}$ and $[K^n(\cdot)X]_t$ into H are Hilbert-Schmidt with Hilbert-Schmidt norms

$$\|K^n(\cdot)X\|_{HS} \leq (MC)^n \frac{T^{n/2}}{n!^{\frac{1}{2}}} \|X\|$$

and

$$\|[K^n(\cdot)X]_t\|_{HS} \leq (MC)^n \frac{T^{(n-1)/2}}{((n-1)!)^{\frac{1}{2}}} \|X\|$$

Proof

The proof is by direct calculation of the Hilbert-Schmidt norm. For an operator with kernel $R_{t,s}$, its Hilbert-Schmidt norm is

$$\|R\|_{HS}^2 = \int_0^T \int_0^T \|R_{t,s}\|_{HS}^2 \, ds \, dt$$

Hence

$$\|[K^n(\cdot)X]_t\|_{HS}^2 =$$
$$= \int_0^t \int_0^{s_1} \cdots \int_0^{s_{n-1}}$$
$$\|S(t-s_1)L(\cdot) \cdots S(s_{n-1}-s_n)L(\cdot)X(s_n)\|_{HS}^2 \, ds$$
$$\leq \int_0^t \int_0^{s_1} \cdots \int_0^{s_{n-1}} (MC)^n \|X(s_n)\|^2 \, ds$$
$$\leq (MC)^n t^{n-1} \|X\|^2 / (n-1)!$$

and

$$\|K^n(\cdot)X\|_{HS}^2 = \int_0^T \|[K^n(\cdot)X]_t\|_{HS}^2 \, ds$$
$$= (MC)^n t^n \|X\| / n!$$

Hence the operators are Hilbert-Schmidt and have the desired H.S. norms. $\square$
Note that the only special property needed in the proof above was that

$$\|B(X,\cdot)\|_{HS} \leq C\|X\|$$

for some $C > 0$.

Note that Hilbert-Schmidtness is equivalent to having a physical random variable only in the linear case. The n-linear forms described above are not necessarily PRV's because they are Hilbert-Schmidt.

3.5 Relation to Ito Integrals

It is easiest to see the relation of white noise to Ito integrals when considering polynomials. The Ito integrals will be seen to be orthogonalized polynomials of a Gaussian random process.

Balakrishnan [4] proved the following theorem relating white noise and Ito polynomials.

Theorem 3.5.1 *Let p_n be an n'th order Hilbert-Schmidt polynomial mapping $\mathcal{W}^n$ into $\mathcal{H}$, and let $\bar{p}_n$ be the symetrized version of p_n. $p_n(N)$ is a physical random variable if and only if the series*

$$p_{n,n-2j}(\cdot) = \sum_{i_1=1}^{\infty} \cdots \sum_{i_j=1}^{\infty} \bar{p}_n(\phi_{i_1}, \phi_{i_1}, \cdots, \phi_{i_j}, \phi_{i_j}, \cdot)$$

converges in Hilbert-Schmidt norm for all CON sequences ϕ and each $j \leq \lfloor n/2 \rfloor$. Further, if $k_{n,n-2j}$ is the kernel of $p_{n,n-2j}$, then $p_n(N)$ has the representation

$$p_n(N) = \sum_{j=0}^{\lfloor n/2 \rfloor} \frac{n!}{(n-2j)! 2^j j!} I_{n-2j}(k_{n,n-2j})$$

where I_j is the j'th order Wiener integral. It is possible to express the second moment of $p_n(N)$ as

$$E[\|p_n(N)\|^2] = \sum_{j=0}^{\lfloor n/2 \rfloor} (\frac{n!}{(n-2j)!2^j j!})^2 \|p_{n,n-2j}\|_{HS}^2$$

In this sense white noise integrals are to polynomials as Ito integrals are to Hermite polynomials. Note that we can also write

$$p_{n,n-2j}(N) = \sum_{i=0}^{\lfloor (n-2j)/2 \rfloor} \frac{(n-2j)!}{(n-2j-2i)!2^i i!} I_{n-2j-2i}(k_{n,n-2j-2i})$$

Hence if the $\lfloor n/2 \rfloor + 1$ dimensional vectors X and Y are defined by

$$X_k = p_{n,n-2k+2}(N), \quad Y_k = I_{n-2k+2}(k_{n,n-2k+2})$$

then we can write

$$X = HY$$

where

$$H_{i,j} = \begin{cases} 1 & \text{if } i = j \\ 0 & \text{if } i > j \\ \frac{(n-2i+2)!}{(n-2i-2j+4)!2^{j-1}(j-1)!} & \text{if } j > i \end{cases}$$

Since this matrix H is always invertible, we can convert back and forth from white noise polynomials to Ito integrals when we have a physical random variable.

Finally, we note that because Ito integrals of different orders are orthogonal, it is possible to express the Hilbert space of square integrable physical random variables, $L_2(\Omega)$, as an orthogonal sum of orthogonal subspaces G_n, each spanned by random variables of the form

$$I_n(k)$$

k ranging over all the Hilbert-Schmidt kernels on $\otimes^n \mathcal{W}$. Hence, $\frac{n!}{(n-2j)!2^j j!} I_{n-2j}(k_{n,n-2j})$ is the projection of $p_n(N)$ onto G_{n-2j}.

The solution of an abstract bilinear system can be considered as a sum of the polynomials $K^n(N)V^0$. The polynomials $K^n(N)V^0$ can be shown to be physical random variables if the two series

$$\sum_{j=0}^{\infty} K(\phi_j)^2 X$$

and

$$\sum_{j=0}^{\infty} K(\phi_j)K(N_1)\cdots K(N_p)K(\phi_j)X$$

converge strongly, independent of the CON sequence ϕ_j. If the series above do converge in this sense, then the limits are $K_D X$ and zero respectively, where

$$[K_D X]_t = \int_0^t S_{t-s} \frac{1}{2} DX_s \, ds$$

and

$$Dx = \sum_{j=1}^{\infty} L(e_j)L(e_j)x$$

for any CON sequence $\{e_j\}$ in H_r. Also, if each of the polynomials $K^n(N)V^0$ can be orthogonalized, then V itself can be expressed as the sum of orthogonal polynomials

$$V = \sum_{j=0}^{\infty} I_j(k_j) \tag{3.14}$$

The Ito polynomials $I_j(k_j)$ are precisely the orthogonal polynomials

$$I_j(k_j) = K_{Ito}^j(W)T(\cdot)V_0$$

where

$$[K_{Ito}(W)X]_t = \int_0^t T_{t-s}L(dW_s)X_s$$

where T is the semigroup generated by $A + \frac{1}{2}D$. As was shown in Theorem 3.4.4,

$$\begin{aligned} E[\|K_{Ito}^n(W)X\|^2] &= \|K^n(\cdot)X\|_{HS}^2 \\ &\leq (CT)^n\|X\|^2/n! \end{aligned}$$

for some C, hence the series in Equation 3.14 converges in mean square.

Using this polynomial method, Balakrishnan [4] showed that if the solution to an abstract bilinear system is a physical random variable, then the system has an equivalent Ito representation. The Ito equation corresponding to

$$\dot{V}_t = AV_t + L(N_t)V_t$$

is

$$dV_t = (A + \frac{1}{2}D)V_t dt + L(dW_t)V_t$$

The representations are equivalent in that the limit of the characteristic functions for $V(P_nN)$ is equal to the characteristic function for the Ito version, and the resulting Ito polynomials are the orthogonalized versions of the white noise polynomials.

Now suppose we would like to change the input in the two equations above. In the white noise case we can always substitute $2N$ for N. However, in the Ito equation, if we substitute $2W$ for W it is also necessary to substitute $4D$ for D. Hence the Ito equation cannot be interpreted as the response of a system to an 'input' dW.

Finally, we consider one more example of a bilinear system with white noise input, taken from [4]. Suppose

$$\dot{V}_t = L(N_t)V_t, \; V_0 \in H$$

where $L(\cdot)$ is a Hilbert-Schmidt mapping from H_r into the Hilbert space of Hilbert-Schmidt operators on H. Suppose also that

$$L(N_1)L(N_2) = L(N_2)L(N_1)$$

Then the solution V_t is given by

$$V_t = \exp\{L(\int_0^t N_s \, ds)\}V_0$$

Now $L(\int_0^t N_s\, ds)$ is a Hilbert-Schmidt map of N into the Hilbert-Schmidt operators, so it is a physical random variable. Since V_t is a continuous function of this term, it is also a physical random variable. In terms of symetrized polynomials, we can write

$$V_t = \sum_{n=0}^{\infty} \frac{1}{n!} L(\int_0^t N_s\, ds)^n V_0$$

Denote

$$p_n(N) = L(\int_0^t N_s\, ds)^n V_0$$

Each $p_n(N)$ is a physical random variable, with

$$p_{n,n-2j}(N) = (tD)^j L(\int_0^t N_s\, ds)^{n-2j} V_0$$

where

$$D = \sum_{k=1}^{\infty} L(e_k)^2$$

and

$$
\begin{aligned}
p_n(N) &= \sum_{j=0}^{\lfloor n/2 \rfloor} \frac{n!}{(n-2j)!\,2^j j!} I_{n-2j}(k_{n,n-2j}) \\
&= \sum_{j=0}^{\lfloor n/2 \rfloor} \frac{n!}{(n-2j)!\,2^j j!} (tD)^j I_{n-2j}(k_{n-2j})
\end{aligned}
$$

Now

$$I_n(k_n) = n![K_{Ito}(W)^n V_0]_t$$

where

$$[K_{Ito}(W)f]_t = \int_0^t L(dW_s)f_s$$

Hence

$$
\begin{aligned}
V_t &= \sum_{n=0}^{\infty} \frac{1}{n!} p_n(N) \\
&= \sum_{n=0}^{\infty} \sum_{j=0}^{\lfloor n/2 \rfloor} \frac{1}{2^j j!} (tD)^j [K_{Ito}(W)^{n-2j} V_0]_t \\
&= \sum_{j=0}^{\infty} \frac{1}{j!} (\frac{t}{2}D)^j \sum_{n=2j}^{\infty} [K_{Ito}(W)^{n-2j} V_0]_t \\
&= e^{\frac{t}{2}D} \sum_{n=0}^{\infty} [K_{Ito}(W)^n V_0]_t
\end{aligned}
$$

which is the solution to the Ito equation

$$dV_t = \frac{1}{2} DV_t + L(dW_t)V_t, \quad V_0 \in H$$

3.6   A White Noise Model For Wave Propagation

White noise in bilinear systems has been studied by Balakrishnan [4],[6]. The white noise model for wave propagation used in most of this monograph is described in [7] and is presented below.

The starting point for the white noise model is the forward scattering equation:

$$\dot{V}_t = \frac{i}{2k}\nabla^2 V_t + ikn_{1t}V_t, \quad V_0 \ given$$

Solutions in the space H defined below were considered in [7], and are the principle subject of this monograph as well.

Definition 3.6.1 *Let H be the complex Hilbert space $\mathbf{L}^2(\mathbf{R}^2)$ with the usual inner product*

$$[f,g] = \int_{\mathbf{R}^2} f_\rho \bar{g}_\rho \ d\rho$$

Define $\mathcal{H}$ to be the Hilbert space $\mathbf{L}^2[(0,T); \mathbf{L}^2(\mathbf{R}^2)]$ with inner product

$$[f,g]_{\mathcal{H}} = \int_0^T [f_t, g_t]$$

The subscript on the inner product will be omitted when no ambiguity exists.

The white noise N was defined to take on values in the real separable Hilbert space defined below.

Definition 3.6.2 *Let H_r be the real Hilbert space $L_2(\mathbf{R}^2)$ with the usual inner product and let $\mathcal{W}$ be the real Hilbert space $L_2([0,T]; L_2(\mathbf{R}^2))$ with the usual inner product.*

At this point a rigorous model for wave propagation can be defined. Solutions to the forward scattering equation:

$$\dot{V}_t = \frac{i}{2k}\nabla^2 V_t + ikn_{1t}V_t \ , \quad V_0 \ given$$

can be described by an abstract bilinear system.

Let V_t take on values in the complex Hilbert space $H = L_2(R^2)$. Let A denote the operator

$$Af = \frac{i}{2k}\nabla^2 f$$

with domain $\mathcal{D}(A) = H^2$. Then A is closed and has dense domain in H. Since $A = -A^*$, A not only generates a C_0 semigroup, but generates a continuous unitary group S_t such that

$$\|S_t X\| = \|X\| \ , \quad \forall X \in H$$

Since n_1 is a homogeneous random field with spectral density P it can be represented as:

$$n_{1t}(\rho) = \int_{\mathbf{R}^2} w(\rho - \rho') N(t,\rho') \ d\rho'$$

where,

$$w(\rho) = \int_{\mathbf{R}^2} e^{i[\rho,\lambda]} P^{\frac{1}{2}}(\lambda) \ d\lambda$$

Since $A(0)$, the variance of n_1, is finite, P is an L_1 function, and $P^{\frac{1}{2}}$ is an L_2 function, making w the inverse Fourier transform of an L_2 function, hence $w \in H_r$. Let:

$$B(X,N)(\rho) = ikX(\rho) \int_{\mathbf{R}^2} w(\rho - \rho') N(\rho') \ d\rho'$$

Then

$$\|B(X,N)\| \ \leq \ k\|X\| \ \sup_{\rho} \ |\int_{\mathbf{R}^2} w(\rho - \rho') N(\rho') \ d\rho'| \qquad (3.15)$$

$$\leq \ k\|X\| \cdot \|w\| \cdot \|N\| \qquad (3.16)$$

Note that the continuity conditions in Equations 3.5 and 3.4 are satisfied by B and S_t, therefore by Theorem 3.3.1 there is a unique mild solution to the forward scattering equation in H.

The Hilbert-Schmidt properties of the operators $K^n(\cdot)X$ are verified for solutions in the space $\mathcal{X}$.

Recall the notation

$$B(X,N) = L(N)X = \mathcal{L}_X N = ikX(w * N)$$

Proposition 3.6.3 *For $X \in H$ and $w \in H_r$, $\mathcal{L}_X$ is H.S. with norm*

$$\|\mathcal{L}_X\|_{HS} = k\|w\|\|X\|$$

Proof

The H.S. norm of an operator with scalar kernel $R(t,s)$ is equal to

$$[\int_{\mathbf{R}^2} \int_{\mathbf{R}^2} |R(t,s)|^2 \ dt \ ds]^{\frac{1}{2}}$$

In this case $R(t,s) = ikX(t)w(t-s)$ so

$$\|\mathcal{L}_X\|_{HS} = k\|X\|\|w\|$$

□

The following proposition follows immediately from Theorem 3.4.4 and Proposition 3.6.3.

Proposition 3.6.4 *The polynomials $K^n(\cdot)X$ are Hilbert-Schmidt mappings from $\otimes^n \mathcal{W}$ into $\mathcal{X}$ for every $X \in \mathcal{X}$, with Hilbert-Schmidt norm*

$$\|K^n(\cdot)X\|_{HS} \leq (k\|w\|)^n T^{n/2} \|X\|/(n!)^{\frac{1}{2}}$$

In addition the polynomials $[K^n(\cdot)X]_t$ are Hilbert-Schmidt as mappings into H with norm

$$\|[K^n(\cdot)X]_t\|_{HS} \leq (k\|w\|)^n T^{(n-1)/2} \|X\|/((n-1)!)^{\frac{1}{2}}$$

Proof

Recall the operator defined above:

$$[K(N)X]_t = \int_0^t S(t-s)B(X(s), N(s)) \; ds$$

Then we have $\|S_t\| = 1$ and by Proposition 3.6.3,

$$\|B(Y,\cdot)\|_{HS} = k\|w\|\|Y\|$$

Hence by Theorem 3.4.4 $K^n(\cdot)X$ is Hilbert-Schmidt with the prescribed norm. In addition , the operators $[K^n(\cdot)X]_t$ are also Hilbert-Schmidt with the prescribed norm. $\square$

The question of when V, or the polynomials $K^n(N)V^0$ are actually physical random variables is still open. Balakrishnan [7] showed that under sufficient smoothness conditions the scalar valued polynomial $[(K^2(N)V^0)_t, \psi]$ is a physical random variable for $V_0, \; \psi \in \mathcal{D}(A)$. In [7] the sums

$$\sum_{j=1}^{\infty} [(K(\phi_{i_1}) \cdots K(\phi_j)K(\phi_j) \cdots K(\phi_{i_{n-2}})V^0)_t, \psi]$$

and

$$\sum_{j=1}^{\infty} [(K(\phi_{i_1}) \cdots K(\phi_j)K(\phi_{i_r}) \cdots K(\phi_j) \cdots K(\phi_{i_{n-2}})V^0)_t, \psi]$$

were shown to converge for all complete orthonormal sequences $\{\phi_j\}$ in $\mathcal{W}$ when $V_0, \; \psi \in \mathcal{D}(A)$ and the polynomials are 'smooth', that is

$$[K(\phi_1) \cdots K(\phi_n)V^0]_t \in \mathcal{D}(A)$$

This smoothness assumption is verified in the next section of this chapter for appropriate conditions on the bilinear operator $B(X, N)$.

Recall from the discussion of the relation of white noise and Ito polynomials that it was sufficient for a physical random variable that the orthogonalized polynomials for $[(K^n(P_m N)V^0)_t, \psi]$ were Cauchy in mean square. The result in [7] demonstrates that the orthogonalized polynomials converge for each sample path, and hence represents a major step towards demonstrating that V_t is a physical random variable. It remains to be shown that the orthogonalized polynomials for terms of order greater than 2 converge in probability or in mean square. The smoothness conditions assumed in [7] are very strict, and would greatly restrict the applicability of white noise theory to infinite dimensional bilinear systems should they prove to be necessary.

In Chapter 4 of this monograph, approximate solutions V^l are defined and shown to be physical random variables that converge in mean square as $l \rightarrow \infty$.

3.7 An Ito Differential Equation Model

Dawson and Papanicolaou [16] considered an Ito stochastic differential equation model and used this to derive limiting values for the distribution of the irradiance function $|V|^2$. They considered solutions to the equation:

$$dV_t = (\frac{i}{2k}\nabla^2 V_t - \frac{k^2}{2}A(0)V_t)dt + ikV_t dW_t, V_0 given \tag{3.17}$$

where

$$V_t \in H = L^2(\mathbf{R}^2)$$

and W is an $S'(\mathbf{R}^2)$ valued Wiener process with covariance

$$E[W_t(\rho)W_s(\rho')] = \min(t,s)A(\rho - \rho')$$

The added correction term $-\frac{k^2}{2}A(0)V_t$ seems arbitrary at first glance, and its correctness needs to be verified by calculating the orthogonalized versions of the white noise polynomials. However, this equation does yield the correct values for the moments, and if V can be shown to be a physical random variable, then the correction term will be

$$-\frac{k^2}{2}A(0)V_t = \frac{1}{2}\sum_{j=1}^{\infty}L(e_j)^2 V_t$$

Addition of this correction term can also be justified using the theory of Stratonovich and considering n_1 as the limit of continuous random processes.

Since

$$E\int_0^t f(t)dW_t = 0$$

in the Ito calculus, it is an easy task to derive moment equations from Equation 3.17 without requiring the variational derivatives used in the Furutsu-Novikov formula.

3.8   The Space H^2

Although I am principally interested in solutions in H it is helpful to consider solutions in two other spaces. First I consider the space of functions which are in $\mathcal{D}(A)$ in the model above.

Definition 3.8.1 *Let H^2 be the complex Sobelev space on $\mathbf{R}^2$ with inner product*

$$[f,g]_2 = [f,g] + [f_x, g_x] + [f_y, g_y] + [f_{xx}, g_{xx}] + [f_{xy}, g_{xy}] + [f_{yy}, g_{yy}]$$

Also denote the Hilbert space $L_2([0,T]; H^2)$ as $\mathcal{H}^2$ with the inner product:

$$[f,g]_{\mathcal{H}^2} = \int_0^T [f_t, g_t]_2 \, dt$$

I wish to establish the existence of mild solutions to Equation 3.7 in $\mathcal{H}^2$.

Now let A be a closed linear operator on H^2 with domain H^4 such that:

$$Af = \frac{i}{2k}\nabla^2 f, \ \forall f \in H^4$$

Once again $A = -A^*$, hence A generates a continuous unitary group S_t on H^2.

Again let B be the bilinear function on $H^2 \times H_r$ defined by

$$B(X,N) = ikX(w * N)$$

where $*$ denotes convolution. Assume that w is in the real Sobelev space H^2 on $\mathbf{R}^2$. Then,

$$\|B(X,N)\|_2^2 = k^2 \sum_{p+q\leq 2} \|(\frac{\partial}{\partial x})^p(\frac{\partial}{\partial y})^q X(w * N)\|^2$$

$$= k^2 \sum_{p+q\leq 2} \|\sum_{j=0}^{p}\sum_{l=0}^{q} \binom{p}{j}\binom{q}{l}[(((\frac{\partial}{\partial x})^j(\frac{\partial}{\partial y})^l w) * N](\frac{\partial}{\partial x})^{p-j}(\frac{\partial}{\partial y})^{q-l} X\|^2$$

$$\leq k^2 \sum_{p+q\leq 2} (\sum_{j=0}^{p}\sum_{l=0}^{q} \binom{p}{j}\binom{q}{l}\|(\frac{\partial}{\partial x})^j(\frac{\partial}{\partial y})^l w\|\|N\|\|(\frac{\partial}{\partial x})^{p-j}(\frac{\partial}{\partial y})^{q-l} X\|)^2$$

$$\leq k^2 \|w\|_2^2\|N\|^2\|X\|_2^2 \cdot 36^2 \cdot 9$$

Hence B is a bounded operator on $H^2 \times H_r$ and satisfies the continuity condition for the abstract bilinear system. Hence equation 3.7 has a unique solution in H^2 and the Cauchy problem is well posed in this space. At this point we can state the following theorem.

Theorem 3.8.2 *Let $H = \mathbf{L}^2(\mathbf{R}^2)$ and $\mathcal{H} = \mathbf{L}^2[(0,T); \mathbf{L}^2(\mathbf{R}^2)]$. Suppose the initial condition V_0 satisfies $V_0 \in H^2$. If w is an element of the real Hilbert space H^2 on $\mathbf{R}^2$ then $V_t \in H^2$ for all $t \in [0,T]$.*

Proof
Since the Cauchy problem was well posed in H^2 it follows trivially that V_t depends continuously (in H^2 norm) on the initial condition $V_0 \in H^2$. $\square$

The following useful corrolary can also be shown.

Corrolary 3.8.3 *Let $H = \mathbf{L}^2(\mathbf{R}^2)$ and $\mathcal{H} = \mathbf{L}^2[(0,T); \mathbf{L}^2(\mathbf{R}^2)]$. Let K and $\mathcal{H}^2$ be defined as above. If $N \in \mathcal{W}$, $w \in H^2$ and $X \in \mathcal{H}^2$ then $K(N)X \in \mathcal{H}^2$ and $(K(N)X)_t \in H^2$, $\forall t \in [0,T]$.*

Proof
Let X be an element of $\mathcal{H}^2$. Then,

$$\|(K(N)X)_t\|_2 = \|\int_0^t S(t-s)B(X(s),N(s))\,ds\| \tag{3.18}$$

$$\leq \int_0^t \|B\|\|X(s)\|_2\|N(s)\|\,ds \tag{3.19}$$

$$\leq \|B\|\|X\|_2\|N\| \tag{3.20}$$

Hence,

$$\|K(N)X\| \leq T\|B\|\|X\|_2\|N\|$$

Therefore $K(N)X \in \mathcal{H}^2$ and $(K(N)X)_t \in H^2$. $\square$

Corrolary 3.8.4 *If $V_0 \in H^2$ and $w \in H^2$ then each term of the Volterra series $K^n(N)V^0 \in \mathcal{H}^2$ and $(K^n(N))_t \in H^2 \forall t \in [0,T]$.*

Proof

Since $V_0 \in H^2$ and S_t is unitary on H^2 we can say that

$$\|V_t^0\|_2 = \|S_t V_0\|_2 = \|V_0\|_2$$

Hence,

$$\|V^0\|_{\mathcal{H}^2} = T^{\frac{1}{2}}\|V_0\|_2$$

Hence $K(N)V^0 \in \mathcal{H}^2$ and $(K(N)V^0)_t \in H^2$ by the previous corrolary. By induction on n, $K^n(N)V^0 \in \mathcal{H}^2$ and $(K^n(N)V^0)_t \in H^2$. $\square$

This verifies the smoothness assumptions in Balakrishnan [7]. One conclusion that can be drawn about the smoothness of the solution V_t is that if $V_0 \in H^m$ and $w \in H^n$ then the Cauchy problem is well posed in $H^{\min(m,n)}$ hence $V_t \in H^{\min(n,m)}$, $\forall t \in [0,T]$.

Proposition 3.8.5 *Let B be a continuous bilinear form mapping $H^2 \times H_r$ into H^2. If*

$$B(X,N) = ikX \cdot (w * N)$$

where $w \in H^2$ then $B(X,\cdot)$ is a Hilbert-Schmidt mapping of H_r into H^2 and

$$\|B(X,\cdot)\|_{HS} \le 324k\|w\|_2\|X\|_2$$

Proof

Let $D^{p,q}$ denote $(\frac{\partial}{\partial x})^p (\frac{\partial}{\partial y})^q$. Let $\{e_j\}$ be a CON sequence in H_r. Then

$$
\begin{aligned}
\|B(X,\cdot)\|_{HS}^2 &= \sum_{j=1}^{\infty} \|B(X,e_j)\|_2^2 \\
&= \sum_{j=1}^{\infty} \sum_{p+q\le 2} \|D^{p,q}B(X,e_j)\|^2 \\
&= k^2 \sum_{j=1}^{\infty} \sum_{p+q\le 2} \| \sum_{n=0}^{p} \sum_{l=0}^{q} \binom{p}{n}\binom{q}{l} D^{n,l}X \cdot [(D^{p-n,q-l}w)e_j]\|^2 \\
&\le k^2 \sum_{j=1}^{\infty} \sum_{p+q\le 2} 81\cdot 16 \sum_{n=0}^{p}\sum_{l=0}^{q} \|D^{n,l}X \cdot [(D^{p-n,q-l}w) * e_k]\|^2 \\
&\le k^2 \sum_{p+q\le 22} 81\cdot 16 \sum_{n=0}^{p}\sum_{l=0}^{q} \|D^{n,l}X\|^2 \|D^{p-n,q-l}w\|^2 \\
&\le k^2 9^4 \cdot 16 \|X\|_2^2 \|w\|_2^2
\end{aligned}
$$

Hence $B(X,\cdot)$ is Hilbert-Schmidt, and its H.S. norm is proportional to the H^2 norm of X. $\square$

The following proposition on the polonomials $K^n(N)X$ can now be proven.

Proposition 3.8.6 *Let X be an element of $\mathcal{H}^2$. Define the operators $K^n(\cdot)X$ as above as mappings from $\otimes^n \mathcal{W}$ into $\mathcal{H}^2$. Then $K^n(\cdot)X$ is Hilbert-Schmidt with H.S. norm:*

$$\|K^n(\cdot)X\| \le (324k\|w\|_2)^n T^{n/2} \|X\|_2$$

Also the mapping $[K^n(\cdot)X]_t$ from $\otimes^n \mathcal{W}$ into H^2 is Hilbert-Schmidt with H.S. norm

$$\|[K^n(\cdot)X]_t\| \le (324k\|w\|_2)^n T^{(n-1)/2} \|X\|_2$$

Proof

Once again $\|S_t\| = 1$ and $\|B(Y,\cdot)\|_{HS} \le 324k\|w\|_2$ by Proposition 3.8.5. Hence by Theorem 3.4.4 the operators $K^n(\cdot)X$ and $[K^n(\cdot)X]_t$ are Hilbert-Schmidt maps into $\mathcal{H}^2$ and H respectively with the prescribed Hilbert-Schmidt norms. $\square$

Note that all of the properties that were shown for solutions in H could also be demonstrated for solutions in H^2 with additional assumptions on B and V_0. The feature that will differentiate the treatment in the two spaces is that in H, $L(N) = -L(N)^*$, making $e^{L(N)}$ a unitary operator on H for each $N \in H_r$. This is not the case in H^2, and this will limit us in the next chapter from considering product form solutions in H^2 in a stochastic input context. I will only be able to demonstrate convergence of the product forms in H^2 for each sample path N.

3.9 The Space $\mathcal{F}$

Finally I consider solutions in the Banach space of Fresnel functions. The Fresnel functions actually form a Banach algebra, but this property will not be used until Chapter 6.

Definition 3.9.1 *Let $\mathcal{F}$ be the Banach space of Fresnel functions on $\mathbf{R}^2$, that is functions of the form:*

$$f(\rho) = \int_{\mathbf{R}^2} e^{i[\rho,\lambda]} \, d\mu(\lambda)$$

where μ is a measure on $\mathbf{R}^2$ with bounded variation. In $\mathcal{F}$ the norm is

$$\|f\|_{\mathcal{F}} = \int_{\mathbf{R}^2} d|\mu|$$

Denote by $B_{\mathcal{F}}$ the Banach space of $\mathcal{F}$ valued functions in $L_2([0,T];\mathcal{F})$ with the norm:

$$\|f\|_{B_{\mathcal{F}}} = \left(\int_0^T \|f_t\|_{\mathcal{F}}^2 \, dt \right)^{\frac{1}{2}}$$

The Fresnel functions are of interest because they include both the beam wave case

$$V_0(\rho) = e^{-|\rho|^2/a}$$

and the plane wave $V_0 \equiv 1$. Fresnel functions are also important in the definition of the Feynman - Ito integral which will be seen in Chapter 6.

For f of the form

$$f(\rho) = \int_{\mathbf{R}^2} e^{i[\rho,\lambda]} \, d\mu$$

let

$$(Af)(\rho) = \int_{\mathbf{R}^2} e^{i[\rho,\lambda]} \frac{i}{2k}|\lambda|^2 \, d\mu$$

whenever

$$\int_{\mathbf{R}^2} |\lambda|^2 \, d|\mu| < \infty$$

Then A is a closed densely defined operator on $\mathcal{F}$. Let $R(p, A) = [pI - A]^{-1}$. Then for p real and positive

$$\|R(p, A)f\| = \int_{\mathbf{R}^2} |\frac{1}{p - \frac{i}{2k}|\lambda|^2}| \, d|\mu| \qquad (3.21)$$

$$\leq \frac{1}{p}\|f\| \qquad (3.22)$$

Hence by the Hille - Yosida theorem A generates a contraction semigroup on $\mathcal{F}$. Since $R(p, -A)$ obeys the same condition, $-A$ also generates a contraction semigroup, hence A generates a unitary group on $\mathcal{F}$.

Once again let B be a bilinear form on $\mathcal{F} \times H_r$ defined by

$$B(X, N) = ikX(w * N)$$

where $w \in \mathbf{L}^2(\mathbf{R}^2)$. Due to the convolution properties of L_1 measures,

$$\|fg\| \leq \|f\|\|g\| \, , \, \forall f, g \in \mathcal{F}$$

Hence,

$$\|B(X, N)\| \leq k\|X\|\|w * N\|$$

Now

$$w * N = \hat{w}\hat{N}$$

where $\hat{f}$ is the Fourier transform of f. Hence

$$\|w * N\| = \int_{\mathbf{R}^2} |\hat{w}(\lambda)\hat{N}(\lambda)| \, d\lambda \qquad (3.23)$$

$$\leq \|w\|_{L_2}\|N\|_{L_2} \qquad (3.24)$$

$$= \|w\|_{L_2}\|N\|_{L_2} \qquad (3.25)$$

by the Plancherel theorem. Hence

$$\|B(X, N)\|_{\mathcal{F}} \leq k\|w\|_{L_2}\|X\|_{\mathcal{F}}\|N\|_{L_2}$$

Hence the conditions of Theorem 3.3.1 for the abstract bilinear system are satisfied in $\mathcal{F}$ so the Cauchy problem is well posed for

$$\dot{V}_t = AV_t + B(V_t, N_t) \, , \, V_0 \in \mathcal{F}$$

In $\mathcal{F}$ the operator $B(X, \cdot)$ is not Hilbert-Schmidt, hence there are no Hilbert-Schmidt polynomials $K^n(N)X$ as were found in H and H^2.

Chapter 4

Product Formula Solutions

The solution to the bilinear differential equation:

$$\dot{X}_t = AX_t + B(X_t, N_t) \; , \; X_0 \; given$$

has no closed form solution in general. Only if A and B commute, i.e.

$$AB(X, N) = B(AX, N) \; , \; \forall N$$

and

$$B(B(X, N_1), N_2) = B(B(X, N_2), N_1) \; , \; \forall N_1, N_2, X \tag{4.1}$$

is there a closed form solution.

Since there is no closed form solution, we consider solutions as the limits of series summations, as in the last chapter, or as limits of products of bounded linear operators, or as path integrals. In this chapter I consider solutions to the forward scattering equation 2.19 as limits of products of linear operators applied to the initial condition and refer to this as a 'product form'. The product forms look like:

$$V_t^l = [\prod_{j=1}^{\lfloor lt/T \rfloor} S(T/l) e^{L(N_j)}] \, V_0$$

These product forms can be applied to very general bilinear systems whenever the commutativity condition 4.1 on B is satisfied. In the case of the forward scattering equation 2.19, product forms are equivalent to the phase-screen approach where the beam is assumed to propagate without distortion except at certain points where it is multiplied by a phase-screen that accounts for the accumulated phase distortion.

In this chapter I review the Trotter-Kato product theory, consider convergence of solutions of parabolic equations with weakly convergent coefficients and apply this to the laser propagation problem.

4.1 Review Of Trotter-Kato Theory

The Trotter-Kato product formula, as stated in Reed and Simon [45] is seen in the theorem below.

Theorem 4.1.1 *Let Y be a Banach space and let A and B be closed linear operators on Y with domains $D(A)$ and $D(B)$ dense in Y that generate contraction semigroups $S(t)$ and $T(t)$ on Y. Suppose $(A+B)$ can be closed on dense domain $D(A) \cap D(B)$ and generates a contraction semigroup $M(t)$ on Y. Then for $y \in Y$*

$$M(t)y = \lim_{l \to \infty}[S(t/l)T(t/l)]^l y \qquad (4.2)$$

When considering solutions to

$$\dot{X_t} = AX_t - BX_t , \ X_0 \text{ given}$$

it may be difficult to study properties of the solutions X_t unless closed form solutions are available. Since this is not the case in general, it can be useful to consider product forms such as in Equation 4.2.

4.2 Convergence Of Solutions To Parabolic Equations With Weakly Convergent Coefficients

In this section I consider the convergence of solutions to vector differential equations with weakly convergent coefficients. First I prove a very useful lemma.

Lemma 4.2.1 *Let H, H_r be separable Hilbert space and denote by $\mathcal{H}$ and $\mathcal{W}$ the Hilbert spaces $L_2([0,T]; H)$ and $L_2([0,T]; H_r)$. Define*

$$[K_l(N)f]_t = \int_0^T R_{t,s}^l L(N_s)f_s \ ds$$

for $N \in \mathcal{W}$ as a map from $\mathcal{H}$ to $\mathcal{H}$. Suppose that $R_{t,s}^l$ is uniformly bounded in l, t, and s by M and that $R_{t,s}^l$ converges strongly to $R_{t,s}$ for each t and s, and $R_{t,s}$ is the kernel of the operator

$$[K(N)f]_t = \int_0^T R_{t,s} L(N_s)f_s \ ds$$

Let $L(\cdot)\phi$ be Hilbert-Schmidt for every $\phi \in H$ with norm

$$\|L(\cdot)\phi)\|_{HS} \le M\|\phi\|$$

Also let V^l be a sequence of functions in $\mathcal{H}$ that are uniformly bounded in t and l by M and that converge to V pointwisely. Then the polynomial map $K_l^m(\cdot)V^l$ converges to $K^m(\cdot)V$ in Hilbert-Schmidt norm as a map from $\otimes^m \mathcal{W}$ into H for each t and into $\mathcal{H}$.

Proof
The proof is by induction on m. For $m = 0$ we have $V^l \to V$ in $\mathcal{H}$ and in H for each t. Assume convergence in Hilbert-Schmidt norm for $m - 1$. Then

$$\|[K_l^m(\cdot)V^l - K^m(\cdot)V]_t\|_{HS} \le$$
$$\le \ \|[(K_l(\cdot) - K(\cdot))K^{m-1}(\cdot)V]_t\|_{HS} + \|[K_l(\cdot)(K_l^{m-1}(\cdot)V^l - K^{m-1}(\cdot)V]_t\|_{HS}$$

Now for the second term above,

$$\|[K_l(\cdot)(K_l^{m-1}(\cdot)V^l - K^{m-1}(\cdot)V]_t\|_{HS}^2 =$$

$$= \int_0^T \|R_{t,s}^l L(\cdot)[K_l^{m-1}(\cdot)V^l - K^{m-1}(\cdot)V]_s\|_{HS}^2 \, ds$$

$$\leq \int_0^T M^4 \|[K_l^{m-1}(\cdot)V^l - K^{m-1}(\cdot)V]_s\|_{HS}^2 \, ds$$

$$\leq M^4 \|K_l^{m-1}(\cdot)V^l - K^{m-1}(\cdot)V\|_{HS}^2$$

and this quantity converges to zero by the induction hypothesis. Since it converges uniformly in t the second term also converges to zero as a map into $\mathcal{H}$. For the first term, let $\{e_p\}$ be a CON sequence in H_r, and $\{f_p\}$ be a CON sequence in $\mathcal{W}$. Then the first term can be expressed as,

$$\|[(K_l(\cdot) - K(\cdot))K^{m-1}(\cdot)V]_t\|_{HS}^2 =$$

$$= \sum_{p_1,\cdots p_{m-1}=1}^{\infty} \int_0^T \sum_{p=1}^{\infty} \|[R_{t,s}^l - R_{t,s}]L(e_p)[K(f_{p_1})\cdots K(f_{p_{m-1}})V]_s\|^2 \, ds$$

For every p, $p_1 \cdots p_{m-1}$, s the innermost term goes to zero as $l \to \infty$. It is bounded by the summable-integrable term

$$4M^2 \|L(e_p)[K(f_{p_{m-1}})\cdots K(f_{p_{m-1}})V]_s\|^2 \tag{4.3}$$

hence the entire expression for the first term goes to zero for each t. Since the bound in Equation 4.3 is not dependent on t, the first term is bounded uniformly in t, and the first term also converges to zero in L_2 norm. Hence the first term converges in H.S. norm as a map into $\mathcal{H}$ as well. Hence the polynomial $[K_l^m(\cdot)V_l]_t$ converges to $[K^m(\cdot)V]_t$ in Hilbert-Schmidt norm for each t and in $\mathcal{H}$. □

In the next theorem, I consider the convergence of mild solutions of an abstract bilinear system with weakly converging coefficients and input. This theorem will be the basis for showing the convergence of the product forms for each sample path N.

Theorem 4.2.2 *Let H and $\mathcal{H} = L_2[(0,T);H]$ and $\mathcal{W} = L_2[(0,T);H_r]$ be separable Hilbert spaces, and let V^l be the mild solution to*

$$\dot{V}_t^l = \alpha_t^l A V_t^l + B(V_t^l, N_t^l), \ V_0^l = V_0 \in H$$

where A generates a unitary semigroup $S(t)$ and B is a bounded bilinear operator with $\|B(X,N)\| \leq \|B\|\|X\|\|N\|$ and Hilbert-Schmidt norm $\|B(X,\cdot)\|_{HS} \leq M\|X\|$. If $\alpha^l \to \alpha$ weakly in $L_2[0,T]$ and $N^l \to N$ weakly in $\mathcal{W}$ then $\forall t \in [0,T]$ $V_t^l \to V_t$ strongly in H and $V^l \to V$ strongly in $\mathcal{H}$, where V is the mild solution to:

$$\dot{V}_t = \alpha_t A V_t + B(V_t, N_t) , \ V_0 \in H$$

Proof
Denote

$$a_{t,s}^l = \int_s^t \alpha_\tau^l \, d\tau$$

Suppose $N^l \equiv 0$. Then $V^l_t = S(a^l_{t,0})V_0$ To see this let h_k be a sequence of step functions converging weakly to α^l. If

$$\dot{X}^k_t = h_k(t) A X^k_t \ , \ X_0 = V_0$$

then the mild solution for X^k is

$$X^k_t = S(\int_0^t h_k(s) \ ds) V_0$$

and $\int_0^t h_k(s) \ ds \to \int_0^t a^l_{t,0}$ therefore $X^k_t \to S(\int_0^t \alpha^l_s V_0 \ ds$ since the semigroup is strongly continuous. If $V_0 \in D(A)$ then $X_t = S(a^l_{t,0})V(0)$ satisfies the equation

$$\dot{X}_t = \alpha^l_t X_t \ , \ X_0 = V_0$$

in the L_2 sense.

Now consider a mild solution to our original equation:

$$V^l_t = S(a^l_{t,0})V_0 + \int_0^t S(a^l_{t,s}) B(V^l_s, N^l_s) \ ds$$

Let

$$V^{0,l}_t = S(a^l_{t,0})V_0$$

In vector form, denote the solution by:

$$V^l = V^{0,l} + K_l(N^l)V^l$$

As in Chapter 3. it is possible to express V^l as

$$V^l = \sum_{m=0}^{\infty} K^m_l(N^l)V^{0,l}$$

To show $V^l \to V$ strongly, I consider the Neuman series for both and compare them term by term.

$$V^l - V = \sum_{m=0}^{\infty} K^m_l(N^l)V^{0,l} - K^m(N)V^0 \qquad (4.4)$$

Since

$$\|K^m_l(N^l)V^{0,l} - K^m(N)V^0\| \leq \frac{(T\|B\|\|N^l\|)^m}{(m-1)!\sqrt{2m}}\|V^{0,l}\| + \frac{(T\|B\|\|N^\|)^m}{(m-1)!\sqrt{2m}}\|V^0\|$$

and

$$\|V^{0,l}\|^2 = \|V^0\|^2 = T\|V_0\|^2$$

and N^l is uniformly bounded for all l, it is enough to show that each term of the series in Equation 4.4 converges to zero strongly. Assume that the $m-1$th term converges to zero. Clearly this is true for $m = 1$. For $m > 1$ we have:

$$K^m_l(N^l)V^{0,l} - K^m(N)V^0 =$$
$$= (K^m_l(N^l)V^{0,l} - K^m(N^l)V^0) + (K^m(N^l)V^0 - K^m(N)V^0)$$

Since $K^m(\cdot)V0$ and $[K^m(\cdot)V^0]_t$ are Hilbert-Schmidt polynomials under the conditions of the theorem and N^n converges weakly to N the second term above converges to zero in H for each t and in $\mathcal{X}$. For the first term, we have

$$\|K_l^m(N^l)V^{0,l} - K^m(N^l)V^0\| \le \|N^l\|^m \|K_l^m(\cdot)V^{0,l} - K^m(\cdot)V^0\|_{HS}$$

If we take $L(N)X = B(X,N)$ and $R_{t,s}^l = S(a_{t,s}^l)$, the conditions of Lemma 4.2.1 are satisfied and the Hilbert-Schmidt norm above converges to zero both in H for each t and in $\mathcal{X}$. Since a weakly convergent sequence is uniformly bounded in norm, the RHS in the above equation goes to zero. Hence each polynomial $K_l^m(N^l)V^{0,l}$ converges to $K^m(N)V^l$ pointwisely in H and also in $\mathcal{X}$. Since the Neuman series for V^l is bounded uniformly in norm by a summable series, $V_t^l \to V_t$ strongly in H for each t and $V^l \to V$ in $\mathcal{X}$.$\square$

4.3 Convergence Of Product Forms For Laser Propagation

In this section I apply the result form the previous section to product forms for solutions to the forward scattering equation 2.19.

Definition 4.3.1 *The product form solution for*

$$\dot{V}_t = AV_t + L(N_t)V_t \ , \ V_0 given \ , \ t \in [0,T]$$

where the Cauchy problem for the above abstract bilinear system is well posed, is:

$$V_t^l = \Theta_t^l \prod_{j=1}^{\lfloor lt/T \rfloor} [e^{\int_{I_j} L(N_s) \, ds} S(T/l)]V_0$$

where $I_j = [(j-1)T/l, jT/l]$ *and for* $t \in I_j$:

$$\Theta_t^l = \begin{cases} S(2t - (j-1)T/l) & t \le (j-\frac{1}{2})T/l \\ e^{L(\int_{(j-1)T/l}^{2t-jT/l} N_s \, ds)}S(T/l) & t \ge (j-\frac{1}{2})T/l \end{cases} \tag{4.5}$$

However, V^l can also be seen to be the mild solution of

$$\dot{V}_t = \alpha_t^l AV_t + B(V_t, N_t^l) \ , \ V_0 \in H$$

where for $t \in I_j = [(j-1)T/l, jT/l]$

$$\alpha_t^l = \begin{cases} 2 & t \le (j-\frac{1}{2})T/l \\ 0 & t > (j-\frac{1}{2})T/l \end{cases}$$

and

$$N_t^l = \begin{cases} 2N_{2t-jT/l} & t \ge (j-\frac{1}{2})T/l \\ 0 & t < (j-\frac{1}{2})T/l \end{cases}$$

The values for α^l and N^l are illustrated in Figures 4.1 and 4.2.

Figure 4.1: Typical values of $N^l(\rho)$

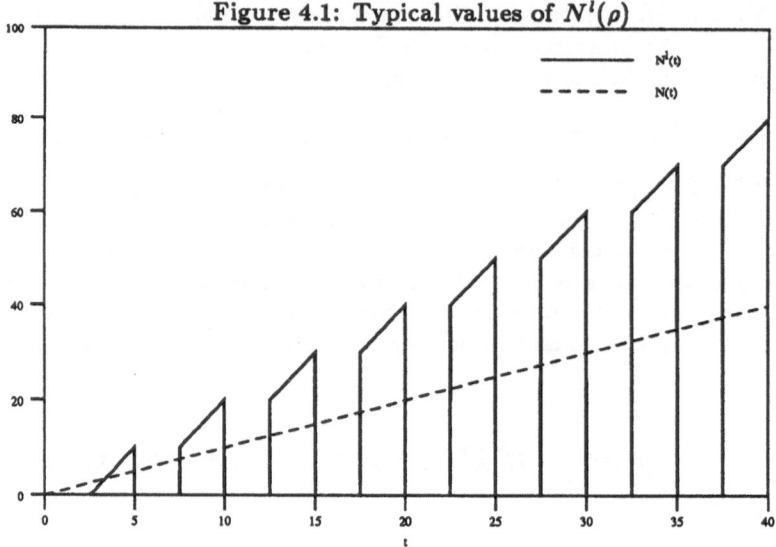

Figure 4.2: Typical values of α^l

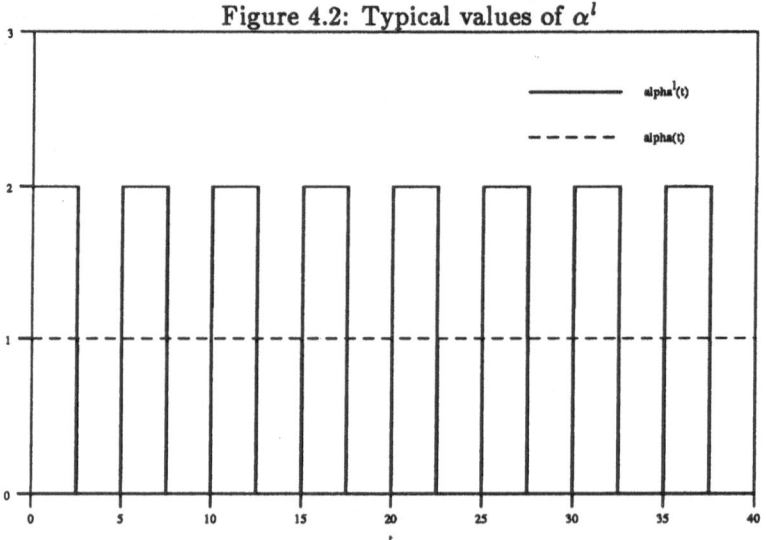

Note that α^l converges to 1 weakly in $L_2[0, T]$. To see this let

$$e_k(t) = \sqrt{2/T} \sin(k\pi t/T)$$

Then e_k is a complete orthonormal sequence and

$$\|[\alpha^l - 1, e_k]\| \leq \sqrt{2/T} k\pi T/l$$

hence $[\alpha^l - 1, e_k] \to 0$ for each k and this is enough to establish weak convergence.

Weak convergence of N^l to N in $\mathcal{W}$ is a little more tricky to show. Consider the $\mathcal{W}$ CON sequence $e_k f_j$ where the f_j are CON in H_r. Then

$$
\begin{aligned}
\|[N^l - N, e_k f_j]\| &= |\int_0^T [N_t^l - N_t, f_j] e_k(t) \, dt| \\
&= |\int_0^T (N_t, f_j)(e_k(\tau_l(t)) - e_k(t) \, dt| \\
&\quad \tau_l(t) \in I_j \text{ when } t \in I_j \\
&\leq \int_0^T |(N_t, f_j)| \sqrt{2/T} k\pi/l \, dt \\
&\leq \|N\| \sqrt{2} k\pi/l
\end{aligned}
$$

Hence $[N^l - N, e_k f_j] \to 0$ for all j and k. Since $e_k f_j$ form a complete orthonormal sequence this is enough to show that $N^l \to N$ in $\mathcal{W}$.

The laser propagation problem, formulated in either $V_t \in H$ or $V_t \in H^2$ with $w \in H^2$ real, satisfied the continuity and Hilbert-Schmidtness conditions set forth in Theorem 4.2.2. Also our generator $A = \frac{i}{2k}\nabla^2$ generates a unitary group in both of these cases. Hence by Theorem 4.2.2 product forms can be used to approximate solutions of the forward scattering equation in H and H^2.

If we restrict α^l and N^l to the forms above it is possible to prove a more general result for the abstract bilinear system.

Theorem 4.3.2 *Let H be a Banach space and let $\mathcal{H} = L_2[(0, T); H]$. Let H_r be a separable Hilbert space and $\mathcal{W} = L_2[(0, T); H_r]$. Let A generate a C_0 semigroup on H, call it S_t, such that $\|S_t\| < M < \infty$ for all $t \in [0, T]$. Let B be a continuous bilinear operator mapping $H \times H_r$ into H such that*

$$\|B(X, N)\| \leq \|B\|\|X\|\|N\|$$

For $N \in \mathcal{W}$ let

$$I_j^l = [(j-1)T/l, jT/l]$$

$$\alpha_t^l = \begin{cases} 2 & t \in I_j^l , \ t \leq (j - \frac{1}{2})T/l \\ 0 & t \in I_j^l , \ t > (j - \frac{1}{2})T/l \end{cases}$$

$$N_t^l = \begin{cases} 0 & t \in I_j^l , \ t \leq (j - \frac{1}{2})T/l \\ 2N(2t - jT/l) & t \in I_j^l , \ t > (j - \frac{1}{2})T/l \end{cases}$$

Then the mild solution of

$$\dot{V}_t^l = \alpha_t^l A V_t^l + B(V_t, N_t^l) , \ V_0^l = V_0 \in H$$

converges to the mild solution of

$$\dot{V}_t = AV_t + B(V_t, N_t) \ , \quad V_0 \in H$$

both pointwisely in H and in H.

Note that H and $\mathcal{H}$ are only Banach spaces here and there is no Hilbert-Schmidtness condition.

Proof

Once again, denote $a_{t,s}^l = \int_s^t \alpha_\tau^l \, d\tau$ and denote $S^l[t, s] = S(a_{t,s}^l)$. Let $V_t^{0,l} = S^l[t, 0]V_0$ and define $g = K_l(N)f$ as an operator from $\mathcal{H}$ into $\mathcal{H}$ as

$$g_t = \int_0^t S^l[t, s]B(f_s, N_s) \, ds$$

As seen in Chapter 3 it is possible to express $V^l - V$ as

$$V^l - V = \sum_{m=0}^{\infty} K_l^m(N^l)V^{0,l} - K(N)^m V^0$$

As in the proof of Theorem 4.2.2, it is enough to show each term of this series converges to zero.

Since $a_{t,s}^l \to 1$ we know $S^l[t, s] \to S_{t-s}$ strongly in H. Since both S^l and S are bounded we know that

$$V_t^{0,l} \to V^0 = S_t V_0$$

both pointwisely in t and in $\mathcal{H}$. Now assume the $m - 1$'th term converges to zero. Then

$$
\begin{aligned}
K_l^m(N^l)V^{0,l} - K^m(N)V^0 &= [K_l(N^l) - K(N)]K^{m-1}(N)V^0 \\
&+ K_l(N^l)[K_l^{m-1}(N^l)V^{0,l} - K^{m-1}(N)V^0] \\
&= [K_l(N^l) - K(N)]f + K_l(N^l)g^l
\end{aligned}
$$

By the induction hypothesis $g^l \to 0$ and since $K_l(N^l)$ is uniformly bounded as a mapping both into H and $\mathcal{H}$ the second term goes to zero in the desired fashion. For the first term,

$$
\begin{aligned}
K_l(N^l)f_t &= \int_0^t S^l[t, s]B(f_s, N_s^l) \, ds \\
&= \int_0^t \tilde{S}_l^l[t, s]B(f_s, N_s^l) \, ds
\end{aligned}
$$

where for $t \in I_j^l$, $s \in I_k^l$

$$
\begin{aligned}
\tilde{S}^l[t, s] &= S^l[t, s] \ , \quad j = k \\
&= S^l[t, (j-1)T/l]S_{(j-k-1)T/l} \ , \quad j > k
\end{aligned}
$$

For the N^l used here S^l and $\tilde{S}^l$ can be used interchangeably in $K^l(N^l)$. Hence,

$$
\begin{aligned}
[K_l(N^l) - K(N)]f_t &= \int_0^t \tilde{S}^l[t, s]B(f_s, N_s^l) - S_{t-s}B(f_s, N_s) \, ds \\
&= \int_0^t [\tilde{S}^l[t, s] - S_{t-s}]B(f_s, N_s) \, ds \\
&+ \int_0^t \tilde{S}^l[t, s]B(f_s, N_s^l - N_s) \, ds
\end{aligned}
$$

In the first integral above, the integrand goes to zero for each s, since $\tilde{S}^l[t,s]$ converges strongly to S_{t-s}. Since the integrand is bounded in norm by

$$2M\|B\|\|f_s\|\|N_s\|$$

the integral converges to zero by the Lebesque Dominated convergence theorem. For $t \in I_j^l$ the second integral can be rewritten as

$$\int_{(j-1)T/l}^{t} \tilde{S}^l[t,s]B(f_s, N_s^l - N_s)\, ds + \int_0^{(j-1)T/l} \tilde{S}^l[t,s]B(f_{\tau(s,l)} - f_s, N_s)\, ds$$

where $|\tau(s,l) - s| \leq T/l$. The first integral above goes to zero since the interval it is taken over goes to zero and the integrand is in L_2. The function f is of the form $K^m(N)V^0$, hence f is uniformly continuous on $[0,T]$. Hence $|f_{\tau(s,l)} - f_s| \to 0$ uniformly in $[0,T]$. Hence the integrand in the second integral goes to zero pointwisely and for l sufficiently large is bounded by $2M\epsilon\|N_s\|$. By the Lebesque Dominated convergence theorem the second integral also goes to zero.

Hence it has been shown that $[K_l^m(N^l)V^{0,l} - K^m(N)V^0]_t$ converges to zero for each t. Since these polynomials are uniformly bounded in both t and l they also converge as functions of t in $\mathcal{H}$. Hence $V^l \to V$ both pointwisely in H and in $\mathcal{H}$. $\square$

Theorem 4.3.3 *Let A generate a C_0 semigroup on the Banach space X and let B be a continuous bilinear operator mapping $X \times H_r$ into X. Suppose that B is commutative in the sense that*

$$B(B(V,N_1),N_2) = B(B(V,N_2),N_1)$$

The abstract bilinear system for A and $B(\cdot,\cdot)$ can approximated by a product form

$$V_t^l = \Theta_t^l \prod_{m=1}^{\lfloor tl/T \rfloor} e^{L(N_{m,l})} S_{T/l}]V_0$$

where

$$N_{m,l} = \int_{(m-1)T/l}^{mT/l} N_s\, ds$$
$$L(N)X = B(X,N)$$

Proof

The product form solution is given by taking V^l as the solution to

$$\dot{V}_t^l = \alpha_t^l A V_t + B(V_t^l, N_t^l)\,, \quad V_0 \in X$$

The rest follows directly from Theorem 4.3.2.$\square$

To specialize to the case of the forward scattering equation with solutions in H, H^2 or $\mathcal{F}$ is trivial at this point.

Theorem 4.3.4 *The solution to the forward scattering equation 2.19 taken in H, H^2 or in $\mathcal{F}$ can be approximated by a product form solution.*

Proof

The formulation of the forward scattering equation 2.19 in all three spaces satisfies the continuity and commutativity conditions in the hypothesis of Theorem 4.3.3, hence the solutions can be approximated by product forms. $\square$

4.4   Product Forms As Physical Random Variables

Since the product forms approximate the solution to the abstract bilinear system, it is desirable to show that they are physical random variables when the input, N, is a $\mathcal{W}$ valued white noise. This is not true in general, and necessary and sufficient conditions are not known. Therefore I consider only the laser propagation problem formulated with $V_t \in H = L_2(\mathbf{R}^2)$ and $V \in \mathcal{X} = L_2[(0,T); H]$. The spaces for the white noise are $N_t \in H_r = L_2(\mathbf{R}^2)$ and $N \in \mathcal{W} = L_2[(0,T); H_r]$ as in Chapter 3. For $A = \frac{i}{2k}\nabla^2$ and $B(X,N) = ikX(w * N)$, $w \in L_2(\mathbf{R}^2)$ we know that the abstract bilinear has a mild solution that is approximated by a product form.

Before presenting the main theorem of this section, I will define some useful notation. Recall that $B(X,N) = L(N)X = \mathcal{L}_X N$. If P_n is a projection operator on $\mathcal{W}$, denote $N^n = P_n N$. For the interval $I_j = [(j-1)T/l, jT/l]$ denote

$$W^{j,n} = \int_{I_j} N_s^n \, ds$$

The l is dropped when the input is just N. The following theorem can be stated.

Theorem 4.4.1 *The product form solution to the abstract bilinear system with A and B for the laser propagation problem above is a physical random variable both at each time t and as a function in $\mathcal{X}$.*

Proof

The product form solution to our problem can be expressed as

$$V_t^l = \Theta_t^l \prod_{j=1}^{\lfloor tl/T \rfloor} [e^{L(W^j)} S_{T/l}] V_0$$

where Θ_t^l is as defined in Equation 4.5. When the input is N^n I will write $V^{l,n}$ as the solution. Let $\{P_n\}$ be a sequence of finite dimensional projection operators on $\mathcal{W}$ such that $P^n \uparrow I$ strongly. Without loss of generality let the range of P_n be spanned by the orthonormal vectors $e_1 \cdots e_n$.

Induction will be used to show that V_t^l is a PRV (physical random variable). Clearly V_0 is a PRV since it is a constant. Suppose that $V_{jT/l}^l$ is a PRV and $V_{jT/l}^l(P_n N)$ is Cauchy in mean square, for all $\{P_n\}$ such that $P_n \uparrow I$ strongly, and $jT/l \le t \le (j+1)T/l$. then

$$\|V_t^{l,n} - V_t^{l,m}\| \le \|[\Theta_t^n - \Theta_t^m]V_{jT/l}^{l,m}\| + \|\Theta_t^n[V_{jT/l}^{l,n} - V_{jT/l}^{l,m}]\|$$

$$\le \|[\Theta_t^n - \Theta_t^m]V_{jT/l}^{l,m}\| + \|V_{jT/l}^{l,n} - V_{jT/l}^{l,m}\|$$

since Θ_t^n is unitary. By our induction hypothesis, the second term goes to zero in probability for $n, m > M$ as $M \to \infty$. For the first term, if $t \le (j + \frac{1}{2})T/l$ then this term is zero and can be ignored. Let $W^n = \int_{jT/l}^{2t-(j+1)T/l} N_s^n \, ds$. Then

$$\|[\Theta_t^n - \Theta_t^m]V_{jT/l}^{l,m}\| = \|[e^{L(W^n)} - e^{L(W^m)}]S_{T/l}V_{jT/l}^{l,m}\|$$

$$= \|[e^{L(W^n - W^m)} - I]S_{T/l}V_{jT/l}^{l,m}\|$$

since $e^{L(N)}$ is unitary for all N. Now

$$E\|[e^{L(W^n - W^m)} - I]S_{T/l}V_{jT/l}^{l,m}\|^2 =$$

$$= E[(2I - e^{L(W^n - W^m)} - e^{L^*(W^n - W^m)})S_{T/l}V_{jT/l}^{l,m}]$$

$$= 2E[(1 - exp\{\frac{1}{2}\sum_{p=m+1}^{n} L^2(\int_{jT/l}^{2t-jT/l} e_p(s)\ ds)\})S_{T/l}V_{jT/l}^{l,m}, S_{T/l}V_{jT/l}^{l,m}]$$

$$= 2E\int_{\mathbf{R}^2}(1 - exp\{-\frac{k^2}{2}\sum_{p=m+1}^{n}(\int_{jT/l}^{2t-jT/l}\int_{\mathbf{R}^2} w(\rho - \rho')e_p(s,\rho')\ d\rho'\ ds)^2\})$$

$$|S_{T/l}V_{jT/l}^{l,m}|_\rho^2\ d\rho$$

Since $V_{jT/l}^{l,m}$ is Cauchy in mean square, we know that $|S_{T/l}V_{jT/l}^{l,m}|^2$ is Cauchy in $L_1(\mathcal{W}, C, \mu)$. This is enough to imply that

$$F_m = E|S_{T/l}V_{jT/l}^{l,m}|^2$$

is Cauchy in $L_1(\mathbf{R}^2)$ and therefore convergent as well. Define the limit to be F. Define

$$g_\rho^{n,m} = 1 - exp\{\frac{-k^2}{2}\sum_{p=m+1}^{n}(\int_{jT/l}^{2t-jT/l}\int_{\mathbf{R}^2} w(\rho - \rho')e_p(s,\rho')\ d\rho\ ds)^2\} \qquad (4.6)$$

and

$$g^M(\rho) = \sup_{n,m>M} g^{n,m}(\rho)$$

For each ρ the function $1_{[jT/l, 2t-jT/l]}w(\rho - \cdot)$ is in $\mathcal{W}$ hence $g^M(\rho) \to 0$ pointwisely. For $n, m \geq M$ the expression above is then

$$\int_{\mathbf{R}^2} g^{n,m}(\rho)F_m(\rho)\ d\rho = \int_{\mathbf{R}^2} g^{n,m}(\rho)[F_m(\rho) - F(\rho)] + g^{n,m}(\rho)F(\rho)\ d\rho$$

$$\leq \int_{\mathbf{R}^2} |F_m(\rho) - F(\rho)| + g^M(\rho)F(\rho)\ d\rho$$

The integral of the first term goes to zero as $M \to \infty, m \geq M$ since F_m goes to F in L_1. The integral of the second term goes to zero since it goes to zero pointwisely and is bounded by the L_1 function F. Hence,

$$\sup_{m,n\geq M} E\|[\Theta_t^n - \Theta_t^m]S_{jT/l}V_{jT/l}^{l,m}\|^2$$

converges to zero as $M \to \infty$, so this term is Cauchy in mean square and hence in probability, since the random variables are uniformly bounded in norm by a constant. Hence $V_t^{l,m}$ is Cauchy in probability and V_t^l is a physical random variable. By induction on j this holds for all $t \in [0, T]$.

Since $\|V_t^{l,m}\| = \|V_0\|$ for all t and m it is also true that $V^{l,m}$ is Cauchy in mean square as an element of $\mathcal{H}$. $\square$

There were three key properties that were used in the proof of Theorem 4.4.1 in addition to the continuity of S_t and Hilbert-Schmidtness of $B(X, \cdot)$. First, $L(N_1)L(N_2) = L(N_2)L(N_1)$. This allowed the use of the exponential $e^{L(N_j)}$ when $\alpha_t^l = 0$. Second, $L(N) = -L(N)^*$ making $e^{L(N_j)}$ a unitary operator. Finally, for any CON sequence $\{\phi_j\}$

in H_r,

$$\sum_{j=1}^{m} L^2(\phi_j)X \to DX$$

strongly in H for each $X \in H$ and D is a bounded operator. In this case $DX = -k^2\|w\|^2$. For any system with these properties, the product form solutions will be physical random variables.

4.5 Convergence Of The Corresponding Ito Integrals

Since the product forms are PRV's they also have a representation in terms of Ito integrals. In this section I show that the Ito integrals corresponding to each product form converge in mean square to the solution of an Ito integral.

Theorem 4.5.1 *The solution to the bilinear system*

$$\dot{X}_t = L(N_t)X_t \ , \ X_0 \in H$$

with

$$L(N)X = ik(w * N)X$$

is a physical random variable, and the corresponding Ito equation for X is:

$$dX_t = -\frac{1}{2}k^2 A(0)X_t \ dt + L(dW_t)X_t \tag{4.7}$$

where $A(0) = \|w\|^2$.

Proof
That X_t and X are physical random variables is clear from Theorem 4.4.1. We need only show that the solution to the Ito equation is the correct one. We note first that

$$-k^2 A(0)X = DX = \sum_{k=1}^{\infty} L^2(e_k)X$$

where $\{e_k\}$ is any CON set in H_r. let

$$p_n(N) = [ikL(\int_0^t N_s \ ds)]^n X_0$$

Then

$$X_t = \sum_{n=0}^{\infty} \frac{1}{n!}p_n(N)$$

Suppose the sequences $\{f_k\}$ and $\{g_m\}$ are CON in $L_2[0,T]$ and H_r respectively. Then $\{f_k g_m\}$ is a CON sequence in $\mathcal{W}$. The summation

$$\sum_{k=1}^{\infty}\sum_{m=1}^{\infty}(L(\int_0^t f_k(s)g_m))^2 X_0$$

converges strongly to $-k^2 A(0)tX_0$, hence in the language of Chapter 3,

$$p_{n,n-2j}(N) = (-k^2 A(0)t)^j p_{n-2j}(N)$$

and

$$p_{n,n-2j}(N) = (\int_0^t)^{n-2j}(-k^2 A(0)t)^j L(N(t_1)) \cdots L(N(t_{n-2j}))X_0 \, dt$$

Let k_n and $k_{n,n-2j}$ denote the kernels of p_n and p_{n-2j} respectively. Then we can write the orthogonal decomposition of p_n as

$$
\begin{aligned}
p^n(N) &= \sum_{j=0}^{\lfloor n/2 \rfloor} \frac{n!}{(n-2j)!2^j j!} I_{n-2j}(k_{n,n-2j}) \\
&= \sum_{j=0}^{\lfloor n/2 \rfloor} \frac{n!}{(n-2j)!2^j j!}(-k^2 A(0)t)^j I_{n-2j}(k_{n-2j}) \\
&= \sum_{j=0}^{\lfloor n/2 \rfloor} \frac{n!}{2^j j!}(-k^2 A(0)t)^j [K_{Ito}(W)^{n-2j}X(0)]_t
\end{aligned}
$$

since

$$I_n(k_n) = n![K_{Ito}(W)^n X_0]_t$$

where

$$[K_{Ito}(W)f]_t = \int_0^t L(dW_s)f_s$$

Then, as we saw in Chapter 3,

$$
\begin{aligned}
X_t &= \sum_{n=0}^{\infty} \frac{p_n(N)}{n!} \\
&= e^{-\frac{k^2}{2}A(0)t} \sum_{n=0}^{\infty} [K_{Ito}(W)^n X_0]_t
\end{aligned}
$$

which is the solution to the Ito equation

$$dX_t = -\frac{k^2}{2}A(0)X_t \, dt + L(dW_t)X_t, \quad X_0 in H$$

□

In developing the Ito version of the product forms it is helpful to note that for half of the time our system behaves like the one in Equation 4.7. The other half of the time it behaves like the unperturbed system $\dot{V} = AV$. Partition the interval into 2^l intervals $I_j^l = [(j-1)T/2^l, jT/2^l]$. Also for $t \in I_j^l$ denote:

$$mid(t,l) = (j - \frac{1}{2})T/2^l$$

$$\lfloor t \rfloor = (j-1)T/2^l$$

$$\lceil t \rceil = jT/2^l$$

$$\tau(t,l) = (t + \lceil t \rceil)/2$$

$$\gamma(t,l) = \begin{cases} \lfloor t \rfloor & t < mid(t,l) \\ 2t - \lceil t \rceil & t \geq mid(t,l) \end{cases}$$

Note that $\tau(t,l)$ and $\gamma(t,l)$ both converge to t uniformly as $l \to \infty$. The equivalent Ito representation of our system can now be defined. For $t \in I_j^l$ let

$$\alpha_t^l = \begin{cases} 2 & t \leq mid(t,l) \\ 0 & \text{otherwise} \end{cases}$$

$$\beta_t^l = 2 - \alpha_t^l$$

$$W^l(t) = W(\gamma(t,l))$$

that is

$$W_t^l = \begin{cases} W_{(j-1)T/2^l} & t \leq mid(t,l) \\ W_{2t-jT/2^l} & t > mid(t,l) \end{cases}$$

where W is an $\mathcal{S}'$ valued generalized Wiener process with covariance

$$E[(W_t, f)(W_s, g)] = \min(t,s)(f,g)$$

for all $f, g \in \mathcal{S}$. The Ito equation corresponding to the product form is

$$dV_t^l = (\alpha_t^l A V_t^l - \beta_t^l \frac{1}{2} k^2 A(0) V_t^l) dt + L(dW_t^l) V_t^l$$

where in this case,

$$-k^2 A(0) X = \sum_{j=1}^{\infty} L^2(e_j) X$$

Define

$$T^l[t,s] = S(\int_s^t \alpha_\tau^l \, d\tau) e^{-\frac{1}{2} k^2 A(0) \int_s^t \beta_\tau^l \, d\tau}$$

Then the mild solution of the Ito equation satisfies

$$V_t^l = T^l[t,0] V_0 + \int_0^t T^l[t,s] L(dW_s^l) V_s^l \,, \quad V_0 \in H \tag{4.8}$$

We can now consider the convergence of the product forms V^l.

Theorem 4.5.2 *The Ito versions of the product forms V^l converge in mean square to the mild solution of*

$$dV_t = (AV_t - \frac{k^2}{2} A(0) V_t) dt + L(dW_t) V_t$$

in mean square.

Proof
If we let

$$[K_l(W^l)f]_t = \int_0^t T^l[t,s] L(dW_s^l) f_s$$

and

$$V_t^{0,l} = T^l[t,0] V_0$$

then the mild solution V^l can be written as

$$V^l = \sum_{m=0}^{\infty} K_l^m(W^l)V^{0,l}$$

the series converging in mean square. Once again it is possible to rewrite K_l as

$$[K_l(W^l)f]_t = \int_0^{\gamma(t,l)} T^l[t,\tau(s,l)]L(dW_s)f_{\tau(s,l)}$$

Let

$$\tilde{T}^l[s,t] = T^l[t,\tau(s,l)]$$
$$\tilde{V}^{0,l}(t) = T^l[\tau(t,l),0]V_0$$

Since $\gamma(\tau(s,l),l) = s$ it is possible to write $K_l^m(W^l)V^{0,l}$ as

$$[K_l^m(W^l)V^{0,l}]_t =$$
$$= \int_0^{\gamma(t,l)} \int_0^{s_1} \cdots \int_0^{s_{m-1}} \tilde{T}^l[t,s_1]L(dW(s_1))\tilde{T}^l[s_1,s_2]L(dW(s_2)) \cdots$$
$$\tilde{T}^l[s_{m-1},s_m]\tilde{V}^{0,l}(s_m)L(dW(s_m))$$

Let

$$T(t)X = S(t)e^{-\frac{k^2}{2}A(0)t}X$$

and define

$$[K(W)f]_t = \int_0^t T(t-s)f_s L(dW_s)$$
$$V^0(t) = T(t)V_0$$

Then following Balakrishnan [4]

$$E\|[K_l^m(W^l)V^{0,l} - K^m(W)V^0]_t\|^2 =$$
$$= \|\int_0^{\gamma(t,l)} \int_0^{s_1} \cdots \int_0^{s_{m-1}}$$
$$[\tilde{T}^l[t,s_1]L(\cdot) \cdots \tilde{T}^l[s_{m-1},s_m]L(\cdot)\tilde{V}^{0,l}(s_m)$$
$$-T(t-s_1)L(\cdot) \cdots T(s_{m-1}-s_m)L(\cdot)V^0(s_m)]\,ds_1 \cdots ds_m\|_{HS}^2$$
$$+(t^m - \gamma(t,l)^m)(k^2 A(0))^m e^{-k^2 A(0)t}/m!$$

Now the first term in the equation above goes to zero by Lemma 4.2.1 since $\tilde{T}^l[t,s]$ converges to $T(t-s)$ strongly and is uniformly bounded, and $\tilde{V}^{0,l}$ converges to V^0 pointwisely and is uniformly bounded by $\|V_0\|$. The second term above goes to zero since $\gamma(t,l)$ converges to t uniformly as l goes to infinity. Hence each of the polynomial terms converges in mean square for each t. Since

$$E\|[K_l^m(W^l)V^{0,l}]_t\|^2 \leq (k^2 A(0)t)^m\|V_0\|^2/m!$$

this is enough to imply that the summation also converges in mean square, hence V_t^l converges to V_t in mean square, where V_t is the solution to:

$$V_t = T(t)V_0 + \int_0^t T(t-s)L(dW_s)V_s$$

Since $\|V_t^l\| = \|V_t\| = \|V_0\|$ almost surely (Dawson and Papanicolaou [16]). This implies that V^l converges to V in mean square as an $\mathcal{H}$ valued random variable. $\square$

This establishes a relation between the white noise and Ito formulations of the laser propagation problem and demonstrates that the correction term $-\frac{k^2}{2}A(0)V_t dt$ is not arbitrary. It is not difficult to see that the limiting Ito representation obtained above is equivalent to that used by Dawson and Papanicolaou [16].

It is important to note that the proof of Theorem 4.5.2 did not rely on any of the unitary properties of $e^{L(N)}$ or S_t, hence the result will be true whenever $L(N)$ satisfies the necessary Hilbert-Schmidtness, and commutativity properties.

Chapter 5

Simulation

Many questions of interest concerning laser distortion cannot be answered analytically, such as the validity of the Markov approximation and the distribution of the irradiance function $|V|^2$. It is also interesting to observe the behavior of the beam as it passes through the atmosphere, which is difficult to observe from an experiment. Therefore digital simulation is a promising tool for studying laser beam distortion.

In this chapter I present simulation strategies, both for solving the forward scattering equation and for generating sample random turbulence fields. The simulations are repeated in a Monte-Carlo fashion to obtain statistics for the mutual coherence function, or second moment, and the distribution of the irradiance function. I then examine simulated solutions using a limiting Ornstein-Uhlenbeck process instead of the white noise process of the Markov approximation.

5.1   Simulation Problem Statement

I wish to simulate solutions of the forward scattering equation

$$\dot{V}_t = \frac{i}{2k}\nabla^2 V_t + ikn_{1t}V_t \, , \ V_0 \in L_2(\mathbf{R}^2)$$

where n_1 is a zero mean Gaussian white noise with covariance

$$E[\int_0^t n_{1r}(\rho_2) \, d\tau \int_0^s n_{1r}(\rho_2) \, d\tau] = \min(t,s)A(\rho_1 - \rho_2)$$

and

$$A(\rho) = 2\pi \int_{\mathbf{R}^2} \Phi(\lambda)e^{i[\lambda,\rho]} \, d\lambda$$

where

$$\Phi(\lambda) = \frac{.033}{4}C_n^2 \frac{e^{-|\lambda|^2/\lambda_m^2}}{(\lambda_0^2 + |\lambda|^2)^{11/6}}$$

is the modified Von Karman spectrum, and

$$\lambda_0 = 2\pi/L_0$$

$$\lambda_m = 2\pi/l_0$$

where L_0 is the outer scale and l_0 is the inner scale. L_0 and l_0 are taken to be

$$l_0 = .001m$$

$$L_0 = 1m$$

The structure function constant C_n^2 is taken to be

$$C_n^2 = 10^{-13}m^{-2/3}$$

which represents strong turbulence. The wave number k is taken to be

$$k = 10^7$$

which is approximately that for visible light. The propagation distance of interest is 1000 . The initial condition V_0 was that for a Gaussian beam and was taken to be

$$V_0(\rho) = e^{-|\rho|^2/10^{-4}}$$

which is a beam with a radius of about .01 m at the near field.

5.2   Application Of Product Formulas

In Chapter 4. I showed that the solutions of bilinear equations could be approximated by product forms. These product forms are easy to produce on a digital computer. They correspond to time-discretized equations of the form:

$$V_{n+1} = e^{ikn_{1,n}}S(T/l)V_n \, , \, V_0 = V(0)$$

where

$$n_{1,n} = \int_{nT/l}^{(n+1)T/l} n_1(t) \, dt$$

$$V_n = V(nT/l)$$

and $S(t)$ is the semigroup generated by $\frac{i}{2k}\nabla^2$.

For solution on a digital computer the spatial dimensions of the solution, that is (x, y) must also be discretized. For this a finite difference method was used to calculate $S(T/l)$. An alternating direction implicit (ADI) method was used for this along with the higher accuracy Douglas formula. This method was first used by Mitchell and Fairweather [36]. An excellent discussion of finite difference methods can be found in Mitchell and Griffiths [37]. The calculations involved are described below.

For each time step, that is step in the direction of propagation, the complex valued function V is sampled on a 33×33 rectangular grid with sampling periods of $\triangle x$ and $\triangle y$ respectively. The outermost samples represent the imposition of boundary conditions. The boundary conditions are taken to be those for the undistorted beam and are approximately zero. The time step is $\triangle t$. Let

$$r = i\triangle t/(\triangle x \cdot \triangle y \cdot 2k)$$

$$(\delta_x^2 V)_{n,m} = V_{n-1,m} - 2V_{n,m} + V_{n+1,m}$$

$$(\delta_y^2 V)_{n,m} = V_{n,m-1} - 2V_{n,m} + V_{n,m+1}$$

The finite difference equations for $S_{\Delta t}$ are then

$$[1 - \frac{1}{2}(r - \frac{1}{6})\delta_y^2]V_{n+1}^* \;=\; [1 + \frac{1}{2}(r + \frac{1}{6})\delta_x^2]V_n \qquad (5.1)$$

$$[1 - \frac{1}{2}(r - \frac{1}{6})\delta_x^2]V_{n+1} \;=\; [1 + \frac{1}{2}(r + \frac{1}{6})\delta_y^2]V_{n+1}^* \qquad (5.2)$$

where V_n^* is an intermediate value with no physical meaning. For maximum accuracy, intermediate boundary values must be selected. For the difference formulas I used the intermediate boundary values are taken to be

$$V_{n+1}^* \;=\; [(r - \frac{1}{6})/(2r)][1 - \frac{1}{2}(r - \frac{1}{6})\delta_y^2]V_{n+1} \qquad (5.3)$$

$$+[(r + \frac{1}{6})/(2r)][1 + \frac{1}{2}(r + \frac{1}{6})\delta_y^2]V_n \qquad (5.4)$$

The right hand side operations, $[1 + \frac{1}{2}(r + \frac{1}{6})\delta^2]$ are calculated directly, or explicitly. The left hand side operations, $[1 - \frac{1}{2}(r - \frac{1}{6})\delta^2]$, are calculated implicitly by inverting a tridiagonal matrix.

In the simulation described here the following numerical values were used.

$$\Delta x = .003\ m$$

$$\Delta y = .003\ m$$

$$\Delta t = 5.0\ m$$

$$k = 10^7$$

$$V_0(x,y) = e^{-10^4(x^2+y^2)}$$

Hence we have a beam with radius 1 cm and are operating in the region of visible light.

Finally, it is possible to improve on the Trotter-Kato product formula here with a minimal increase in computation time. Instead of

$$V_{n+1} = e^{ikn_{1,n}} S_{\Delta t} V_n$$

I use

$$V_{n+1} = e^{\frac{1}{2}ikn_{1,n}} S_{\Delta t} e^{\frac{1}{2}ikn_{1,n}} V_n$$

The improvement in accuracy occurs because the value of r is small, $r = i/36$, relative to the typical values for $ikn_{1,n}$ which are generally on the order of plus or minus $.7i$ or so. This is discussed heuristically below.

Let A and B be matrices and assume that the largest values in A are several times larger than the largest values in B. In Table 5.1 the expansions of $e^A e^B$ and e^{A+B} are compared up to the first four terms.

It is easy to see from table 5.1 that the straightforward product formula is only accurate in the first two terms when A and B don't commute. I now consider two modifications that represent little increase in computation time. First I consider approximating e^{A+B} by $\frac{1}{2}(e^A e^B + e^B e^A)$. As can be seen from table 5.2, this method is

Term	e^{A+B}	$e^A e^B$
0	I	I
1	$A + B$	$A + B$
2	$\frac{1}{2}(A^2 + AB + BA + B^2)$	$\frac{1}{2}(A^2 + 2AB + B^2)$
3	$\frac{1}{6}(A^3 + A^2B + ABA + BA^2)$ $+\frac{1}{6}(AB^2 + BAB + B^2A + B^3)$	$\frac{1}{6}(A^3 + 3A^2B)$ $+\frac{1}{6}(3AB^2 + B^3)$
4	$\frac{1}{24}(A^4 + B^4)$ $+\frac{1}{24}(A^3B + A^2BA + ABA^2 + BA^3)$ $+\frac{1}{24}(A^2B^2 + ABAB + AB^2A)$ $+\frac{1}{24}(BABA + B^2A^2 + BA^2B)$ $+\frac{1}{24}(B^3A + B^2AB + BAB^2 + AB^3)$	$\frac{1}{24}(A^4 + B^4)$ $+\frac{1}{24}(4A^3B)$ $+\frac{1}{24}(6A^2B^2)$ $+\frac{1}{24}(4AB^3)$

Table 5.1: Comparison of e^{A+B} and $e^A e^B$

Term	$\frac{1}{2}(e^A e^B + e^B e^A)$	$\frac{1}{2}(e^A e^B + e^B e^A) - e^{A+B}$
0	I	0
1	$A + B$	0
2	$\frac{1}{2}(A + B)^2$	0
3	$\frac{1}{6}A^3 + \frac{1}{4}(A^2B + BA^2)$ $+\frac{1}{4}(AB^2 + B^2A) + \frac{1}{6}B^3$	$\frac{1}{12}(A^2B - 2ABA + BA^2)$ $+\frac{1}{12}(B^2A - 2BAB + AB^2)$
4	$\frac{1}{24}(A^4 + 2A^3B + 2BA^3)$ $+\frac{1}{8}(A^2B^2 + B^2A^2)$ $+\frac{1}{12}(AB^3 + B^3A) + \frac{1}{24}B^4$	$\frac{1}{24}(A^3B + BA^3 - A^2BA - ABA^2)$ $+\frac{1}{24}(2A^2B^2 + 2B^2A^2 - AB^2A)$ $-\frac{1}{24}(ABAB + BABA + BA^2B)$ $+\frac{1}{24}(AB^3 + B^3A - B^2AB - BAB^2)$

Table 5.2: Comparison of $\frac{1}{2}(e^A e^B + e^B e^A)$ and e^{A+B}

Term	$e^{A/2}e^{B}e^{A/2}$	$e^{A/2}e^{B}e^{A/2} - e^{A+B}$
0	I	0
1	$A + B$	0
2	$\frac{1}{2}(A+B)^2$	0
3	$\frac{1}{6}(A^3 + B^3) + \frac{1}{4}(AB^2 + B^2 A)$ $+\frac{1}{8}(A^2 B + BA^2 + 2ABA)$	$\frac{1}{12}(AB^2 + B^2 A - 2BAB)$ $+\frac{1}{24}(2ABA - A^2 B - BA^2)$
4	$\frac{1}{24}A^4 + \frac{1}{16}ABA^2$ $+\frac{1}{48}(A^3 B + BA^3 + 3A^2 BA)$ $+\frac{1}{16}(A^2 B^2 + 2AB^2 A + B^2 A^2)$ $+\frac{1}{12}(AB^3 + B^3 A) + \frac{1}{24}B^4$	$\frac{1}{48}(A^2 BA + ABA^2 - A^3 B - BA^3)$ $+\frac{1}{48}(A^2 B^2 + B^2 A^2 + 4AB^2 A)$ $-\frac{1}{24}(ABAB + BABA + BA^2 B)$ $+\frac{1}{24}(AB^3 + B^3 A - B^2 AB - BAB^2)$

Table 5.3: Comparison of $e^{A/2}e^{B}e^{A/2}$ and e^{A+B}

a definite improvement. This approximation shows some improvement over using the product formula directly. The first three terms are exact and the error in the other terms is reduced somewhat. Since B is assumed to be smaller than A in terms of the size of its elements, the terms that are first order in B will dominate.

A second modification, that takes advantage of the relative magnitudes of A and B approximates e^{A+B} by $e^{A/2}e^{B}e^{A/2}$. This approximation reduces the error in the terms that are first order in B, as shown in table 5.3. The error is reduced by about a factor of two from the approximation in table 5.2.

5.3 Generating Pseudo-Random Fields

Simulation of laser propagation in a random media requires the computation of a large number of pseudo-random fields for the index of refraction deviations n_1. Recall from Chapter 2 that n_1 is assumed to be an isotropic, homogeneous random field in x and y and uncorrelated in the direction of propagation. Because the spectral density for n_1 is non-rational, it is not possible to generate random fields as the solution to a difference equation with white noise input.

One way to produce the 31×31 random fields is to factor the 961×961 covariance matrix into the form:

$$R = LL^*$$

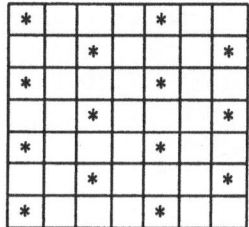

Figure 5.1: Sampling Pattern For Generating Random Turbulence Fields

where R is the covariance of the sampled version of n_1 and L is a lower triangular matrix. The Croot algorithm can be used to compute L fairly easily. The random field is produced by taking

$$n_1 = LN$$

where N is a 961 dimensional white noise vector, i.e $Cov(N) = I$. This is the most accurate, but also the most expensive way to generate the random fields. On the FPS 164 Array Processor this method required 46 seconds to produce 200 random fields.

A somewhat faster method consists of generating 264 samples of the random field using the sampling pattern shown in Figure 5.1, interpolating the unsampled points, and adding a white noise process so that the variance of n_1 is the same at every point. If the variance is taken to be 1.0 at every point, then the maximum interpolation error was 1.06×10^{-3} and the maximum difference between the correct covariance and the one produced using this method is 4.4×10^{-4}. This is also highly accurate and only required 22 seconds to produce 200 random fields.

As can be seen from the sampling pattern, each unsampled point is adjacent to either two or three sampled points. This makes the interpolation very accurate since the covariance function drops off slowly.

The covariance function is found by numerically computing the Fourier transform of the spectral density. Since n_1 is isotropic, the spectral density $\Phi(\lambda)$ is a function of $|\lambda|$ only, hence

$$\int_{\mathbf{R}^2} e^{i[\lambda,\rho]} \Phi(\lambda) \ d\lambda = 2\pi \int_0^\infty J_0(\lambda\rho)\Phi(\lambda)\lambda \ d\lambda$$

In this monograph I use the Modified Von Karman Spectrum

$$\Phi(\lambda) = .033 C_n^2 \frac{e^{-\lambda^2/\lambda_m^2}}{(\lambda_0^2 + \lambda^2)^{11/6}}$$

with parameter values

$$\begin{aligned} C_n^2 &= 10^{-13} m^{-2/3} \\ \lambda_m &= 2\pi/l_0 \\ \lambda_0 &= 2\pi/L_0 \end{aligned}$$

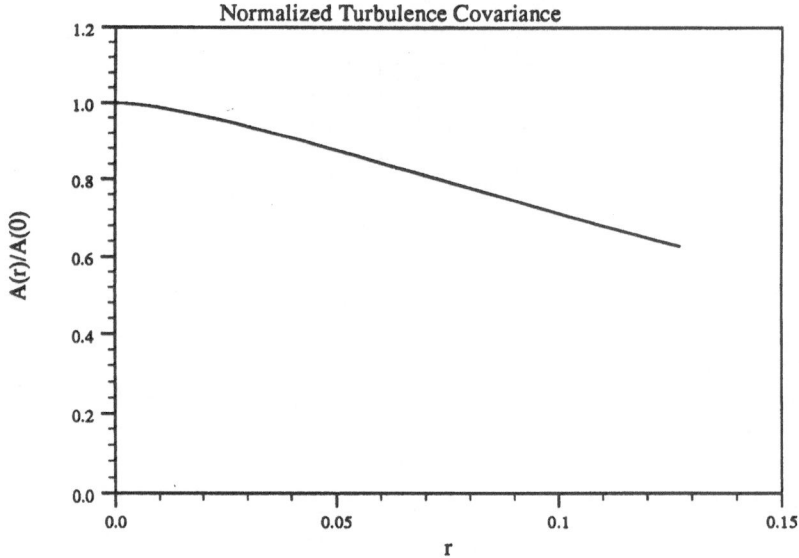

Figure 5.2: Normalized Covariance Function Of n_1

where l_0 and L_0 are the inner and outer scale respectively. Their values are taken to be

$$l_0 = .001m$$
$$L_0 = 1.m$$

The value of $C_n^2 = 10^{-13}m^{-2/3}$ corresponds to very strong turbulence. $C_n^2 = 10^{-14}m^{-2/3}$ would be considered strong turbulence. A graph of the covariance function is shown in Figure 5.2.

5.4 Weak Convergence Of Trigonometric Series

In this section I explore an alternative method for generating random fields for digital simulation. A homogeneous Gaussian random field is approximated in distribution by a random trigonometric series, that is one with random frequencies and phases. This method allows us to generate a continuous waveform on the computer and sample it at arbitrary points. A modification of this method uses predetermined frequencies and random phases to obtain a random field with a particular spectral density.

Let X be a zero mean homogeneous Gaussian random field on $\mathbf{R}^2$, with covariance R and spectral density $\sigma^2 \Phi(\lambda)$ where

$$\int_{\mathbf{R}^2} \Phi(\lambda) \, d\lambda = 1$$

and

$$\int_{\mathbf{R}^2} \Phi(\lambda) |\lambda|^2 \, d\lambda < \infty$$

I wish to approximate X by a sequence of the form:

$$X_\rho^n = \sqrt{\frac{2\sigma^2}{n}} \sum_{k=1} n \cos(\lambda_k \cdot \rho + \theta_k)$$

for $\rho \in [0, T]^2$. The λ_k are iid random variables with probability density function $\Phi(\lambda)$ and the θ_k are iid and are uniformly distributed on $[0, 2\pi]$. Hence,

$$E[X_\rho^n X_{\rho'}^n] = R(\rho - \rho')$$

and

$$E[X_\rho^n] = 0$$

so X and X^n have the same mean and covariance function. If X^n were Gaussian this would be enough to imply that they were identically distributed. The Central Limit Theorem tells us that any finite number of sample points of X^n converges to a Gaussian vector with the same distribution as the same samples of X.

Since a finite number of sample points is all we use in a computer simulation, this is really enough. The method outlined above does have a few drawbacks, however. Since the frequencies λ_k are chosen randomly, a very large number of terms may need to be used to obtain a reasonable sample path. Using predetermined frequencies yields an approximate covariance function for X^n

$$R^n(\rho) = (\sigma^2/n) \sum_{k=1}^{n} \cos(\lambda_k \cdot \rho)$$

The frequencies are then chosen so R^n approximates R.

This method for generating random processes appeared to work well for one dimensional random processes, however for two dimensional random fields, the number of terms needed to obtain good samples and a good approximation to the covariance function was large enough so that the computation time was not much less than that needed for the matrix factorization method described in the previous section. Hence for my particular problem, this was not the method of choice. However this method might be useful in problems requiring arbitrary sampling or continuous waveforms.

For problems where continuous waveforms are needed, and theoretical studies, the construction for X^n is very interesting. Using several results from Billingsley [10], we can show that $X^n \xrightarrow{d} X$ in the space of continuous functions $C[0, T]^2$ which I shall refer to as C. I will present a few definitions and theorems relating to weak convergence of measures from Billingsley [10] and then apply them to the case considered here.

Definition 5.4.1 *A family Π of probability measures is relatively compact if every sequence in Π contains a weakly convergent subsequence converging to a probability measure (not necessarily one in Π).*

Definition 5.4.2 *A family Π of probability measures is tight if for each $\epsilon > 0$ there is a compact set K such that $P(K) > 1 - \epsilon$, $\forall P \in \Pi$.*

The following theorem, due to Prokhorov, equates these two definitions.

Theorem 5.4.3 (Prokhorov) *A family* Π *of probability measures is tight if and only if it is relatively compact.*

The idea is to have a test for tightness of measures on C and use this to show weak convergence of the measures. The next two theorems are found in Billingsley [10].

Theorem 5.4.4 *Let* $\{P_n\}$ *be a sequence of probability measures on* C. *If the finite dimensional distributions of* P_n *converge to the finite dimensional distributions of a probability measure* P *on* C *and* $\{P_n\}$ *is relatively compact, then* P_n *converges weakly to* P.

Definition 5.4.5 *Let*

$$w_x(\delta) = \sup_{|\rho - \rho'| < \delta} |x_\rho - x_{\rho'}|$$

Theorem 5.4.6 *A sequence of probability measures* $\{P_n\}$ *on* C *is tight if and only if for each* ϵ *there exists* a *and* n_0 *such that*

$$P_n[x : |x(0)| > a] < \epsilon, \ n \geq n_0$$

and for each η *and* ϵ *there exist* δ *and* n_0 *such that*

$$P_n[x : w_x(\delta) > \eta] < \epsilon, \ n \geq n_0$$

Theorem 5.4.6 is proved only for $C[0, T]$ in Billingsley, however the result is easily extended to $C[0, T]^d$. Taking Theorems 5.4.3, 5.4.4 and 5.4.6 together, we know that any sequence of measures satisfying the conditions in Theorem 5.4.6 is tight, hence relatively compact by Prokhorov's Theorem. If in addition their finite dimensional distributions converge to the finite dimensional distribution of a probability measure on C, the sequence of measures converges weakly. Hence we have the following theorem.

Theorem 5.4.7 *If a sequence of probability measures* $\{P_n\}$ *on* C *satisfies the conditions in Theorem 5.4.6 and the finite dimensional distributions of* P_n *converge to the finite dimensional distributions of* P, *a probability measure on* C, *then* P_n *converges weakly to* P.

It remains to be shown that the measures corresponding to our trigonometric approximation converge weakly on $C[0, T]^2$. One more result is required to put these conditions in a practical form. This is the multidimensional version of the Kolmogorov Lemma, which can be found in Yadrenko [55].

Proposition 5.4.8 *Let* X *be a random field on* $[0, T]^2$. *If*

$$E[|X_\rho - X_{\rho'}|^\alpha] \leq M|\rho - \rho'|^{2+\beta}$$

for some α, $\beta > 0$ *and all* ρ, $\rho' \in [0, T]^2$ *then* X *is continuous with probability one and*

$$Prob(w_X(\delta) > a) \to 0 \text{ as } \delta \to 0$$

Theorem 5.4.9 *For X and X^n defined above,*

$$X^n \xrightarrow{d} X$$

as a $C[0,T]^2$ valued random variable if X is twice mean square differentiable.

Proof

The finite dimensional distributions of

$$X^n(\rho) = \sqrt{2\sigma^2/n} \sum_{k=1}^{n} \cos(\lambda_k \cdot \rho + \theta_k)$$

converge to those for X because they have the same first two moments and by the Central Limit Theorem. It remains to be shown that the corresponding measures satisfy the conditions in Theorem 5.4.6 and that the limit is a probability measure on C. The first condition is easily shown to be satisfied.

$$Pr(|X^n(0)| > a) \leq E[X^n(0)^2]/a^2 = \sigma^2/a^2$$

by Tchebychev's inequality, and for any ϵ we can choose $a > \sigma/\sqrt{\epsilon}$ to make

$$Pr(|X^n(0)| > a) < \epsilon$$

The continuity condition requires a little more work. To show tightness we only need consider the moments of increments of X^n. For a twice differentiable random field the covariance function can be expressed as

$$R(\rho) = 1 - a(\rho, \rho) - b(\rho)$$

where a is a bilinear form and $|b(\rho)| = O(\rho^4)$. Then

$$
\begin{aligned}
E[(X^n(\rho') - X^n(\rho + \rho'))^4] &= \\
&= \frac{12\sigma^2}{n}[1 - R(\rho) - \frac{1}{4}(1 - R(2\rho)] + 12\sigma^2(1 - \frac{1}{n})[1 - R(\rho)]^2 \\
&= \frac{12\sigma^2}{n}[b(\rho) - \frac{1}{4}b(2\rho)] + 12\sigma^2(1 - \frac{1}{n})[a(\rho, \rho) + b(\rho)]^2 \\
&\leq M|\rho|^4
\end{aligned}
$$

Hence the condition in Proposition 5.4.8 is satisfied and hence the continuity condition is satisfied. By Theorem 5.4.6 this implies that the measures for X^n are tight and hence, by Prokhorov's theorem, relatively compact. Because the finite dimensional distributions converge , this is enough to imply weak convergence. Hence X^n converges in distribution as a $C[0,T]^2$ valued random variable.

We also have that

$$
\begin{aligned}
E[|X_{\rho'+\rho} - X_{\rho'}|^4] &= 12\sigma^2[1 - R(\rho)]^2 \\
&\leq M|\rho|^4
\end{aligned}
$$

Hence the measure for the Gaussian process X is concentrated on $C[0,T]^2$. and $X^n \xrightarrow{d} X$ as a $C[0,T]^2$ valued random variable. $\square$

Our n_1 process with the modified Von Karman spectral density

$$\Phi(\lambda) = \frac{C_n^2 e^{-\lambda^2/\lambda_m^2}}{(\lambda_0^2 + \lambda^2)^{11/6}}$$

is twice mean-square differentiable, so for generating random turbulence fields the trigonometric series approximation converges in distribution as a continuous function.

5.5   The Mutual Coherence Function

The mutual coherence function is the spatial correlation function of the beam intensity at each distance z (t) along the direction of propagation. Thus,

$$\Gamma_t(\rho_1, \rho_2) = E[V_t(\rho_1)\overline{V_t(\rho_2)}] \tag{5.5}$$

where $\overline{V}$ denotes the complex conjugate of V.

Under the Markov approximation, the coherence function can be shown to obey the partial differential equation

$$\frac{\partial}{\partial t}\Gamma_t = \frac{i}{2k}(\nabla_1^2 - \nabla_2^2) - \frac{k^2}{2}D\Gamma$$

where D is the structure function for n_1. In the plane wave case $(V_0 \equiv 1)$ the coherence function can be calculated exactly to be

$$\Gamma_t(\rho_1, \rho_2) = e^{-k^2 D(\rho_1 - \rho_2)/2}$$

In the general case, calculation of Γ is more difficult. One integral form solution is

$$\Gamma_t(\rho, \rho') = (\frac{1}{4\pi^2})^2 \int_{\mathbf{R}^4} e^{-\frac{k^2}{2}\int_0^t D(\rho - \rho' - (\mu - \mu')\tau/k)\ d\tau}$$
$$e^{\frac{-it}{2k}(|\mu|^2 - |\mu'|^2)} e^{i([\mu, \rho] - [\mu', \rho'])} \hat{V}_0(\mu)\overline{\hat{V}_0(\mu')}\ d\mu d\mu'$$

The integral above was computed numerically, and these values are compared with results of a Monte Carlo simulation useing 8000 runs in the discussion below. In particular, the average irradiance and the correlation of the beam intensity at the boresight with its value at other points are considered.

If we wish to operate a laser tracking or communication system with a photocollector receiver, the received signal will be the integral over the aperture of the receiver of the irradiance function $I(\rho) = |V(\rho)|^2$. If this received signal is put through a low pass filter, the result will approximate the *average in time* of the irradiance function integrated over the aperture. If the turbulence field is relatively uncorrelated in time, over intervals on the order of 1 sec, the filter output should approximate the integral of the *expected irradiance* function over the aperture. Note that the integral of the expected irradiance over $\mathbf{R}^2$ should be a constant independent of the turbulence statistics. It is the spreading and 'motion' of part of the beam off of the aperture that results in the loss of received energy. Hence the average irradiance is of potential interest in problems of tracking and communication. The time correlation of the turbulence field is beyond the scope of this monograph and will not be considered further.

The theoretical values, 'theory', for the average irradiance function $\Gamma(\rho, \rho)$ are compared with the simulated averages , 'simulation', for a sample of 8000 Monte Carlo runs and the no turbulence case 'undistorted' in Figure 5.3. The agreement between the simulation and theory is good in general, and this also serves to confirm the accuracy of the simulation.

For the values along the line $(0, x)$, as in Figure 5.3, the effect of the turbulence is to increase the spreading of the average irradiance. For the values along the line $(.021, x)$,

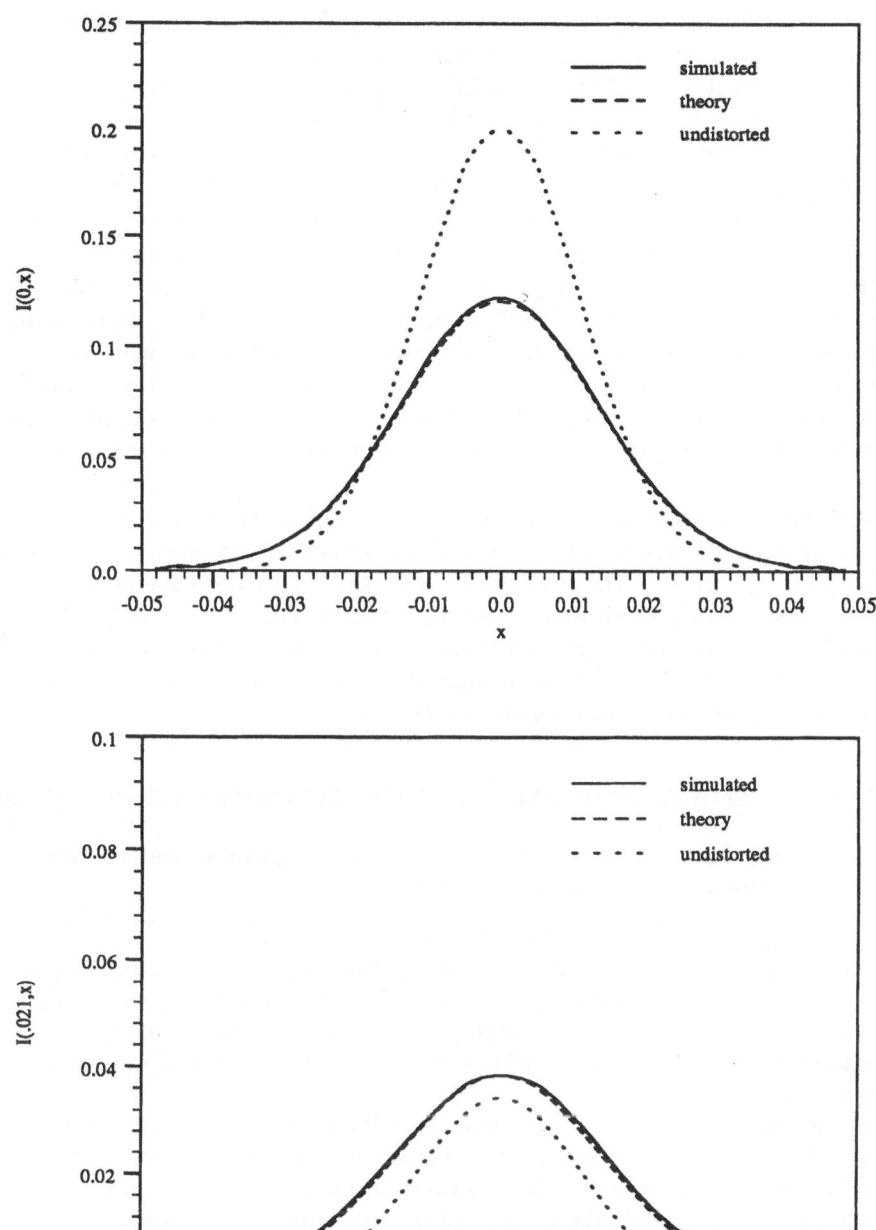

Figure 5.3: Average Irradiance Function Across Boresight, $I(0, x)$, And Off Boresight, $I(.021, x)$, At 1000 m, Based On 8000 Samples.

as in Figure 5.3, the effect of the turbulence is to increase the spreading of the average irradiance, which increases the values at these particular locations.

The turbulence has the effect of causing the average irradiance to spread faster than that for the undistorted beam. This can be somewhat misleading, however, since this 'spreading' of the average may reflect the shifting of the beam in a random manner rather than its spreading. Some simulations actually showed the distorted beam to have a smaller radius and a higher peak than the undistorted beam.

The theoretical values and simulation averages for the coherence function $\Gamma(0,0;0,x)$ and $\Gamma(0,0;.021,x)$ are given in Figures 5.4 and 5.5. I consider the correlation of the beam intensity at the boresight $V(0,0)$ with values taken on a line passing through the boresight and on a line displaced .021 m from the boresight. Once again the agreement between theory and simulation is good. Again, it can be misleading to interpret the attenuation of the average coherence function as an attenuation of the beam due to the turbulence. The bending of the beam, or it shifting in a random manner also accounts for this.

Attenuation of the coherence function may also reflect the distortion of the phase of the beam by random turbulence. On the line through the boresight, Figure 5.4 shows that the primary effect of the turbulence is to attenuate the coherence function and shift the peaks somewhat, without changing the basic shape very much. Along the line displaced from the boresight, as in Figure 5.5 the effect is quite different, indicating that more than a simple loss of phase coherence is occurring. Once again, the bending of the beam may be crucial to interpreting this effect.

5.6   The Distribution Of The Irradiance Function

One of the many unsolved problems of wave propagation in a random medium is the probability distribution of the irradiance function

$$I_\rho = |V_\rho|^2 \tag{5.6}$$

In this section I consider several possible distributions for the irradiance function and fit them to the simulated irradiance data at 1000 m. The irradiance was sampled at the points (in the x,y plane) $(0,0)$, $(0,.021)$, $(0,-.021)$, $(.021,0)$, $(-.021,0)$, $(.021,.021)$, $(.021,-.021)$, $(-.021,.021)$ and $(-.021,-.021)$ after the beam had propagated 1000 m through simulated turbulence.

In the saturation region, which occurs over longer distances (e.g. over 100 m), or in strong turbulence the classical perturbation methods predict poor or contradictory distributions for the irradiance. The method of small perturbations predicts a Rice-Nakagami distribution, while the method of smooth perturbations predicts a log-normal distribution. Both of these methods break down in the saturation region as has been shown theoretically.

Under the method of small perturbations, the solution to the forward scattering equation is

$$V_t \approx S_t V_0 + \int_0^t S_{t-s} L(N_s) S_s V_0$$

which is a Gaussian random vector with non-zero mean when N is a white noise. In Chapter 3 it was shown that for sufficient smoothness conditions on $L(\cdot)$, the solution

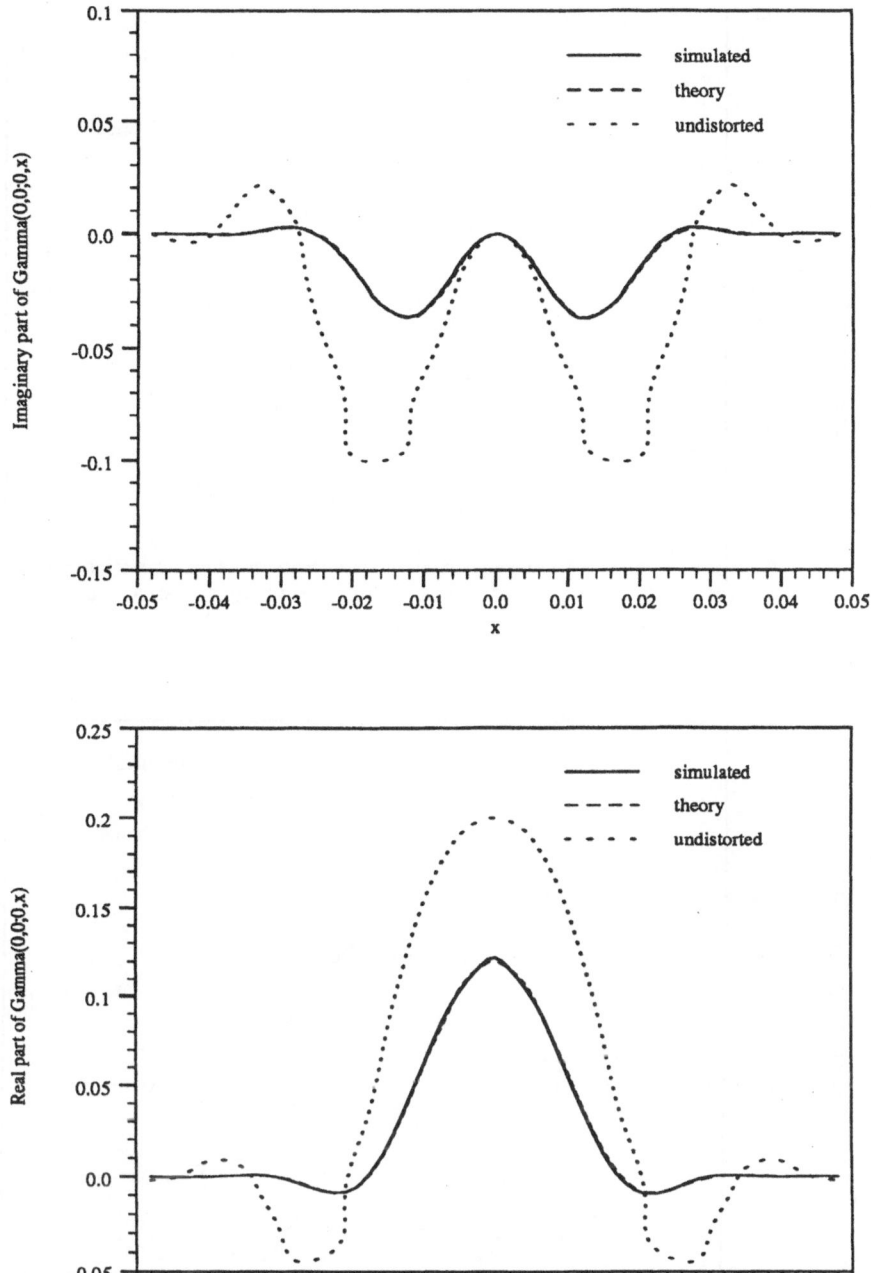

Figure 5.4: **Average Coherence Function Across Boresight, $\Gamma(0,0;0,x)$, At 1000 m, Based On 8000 Samples**

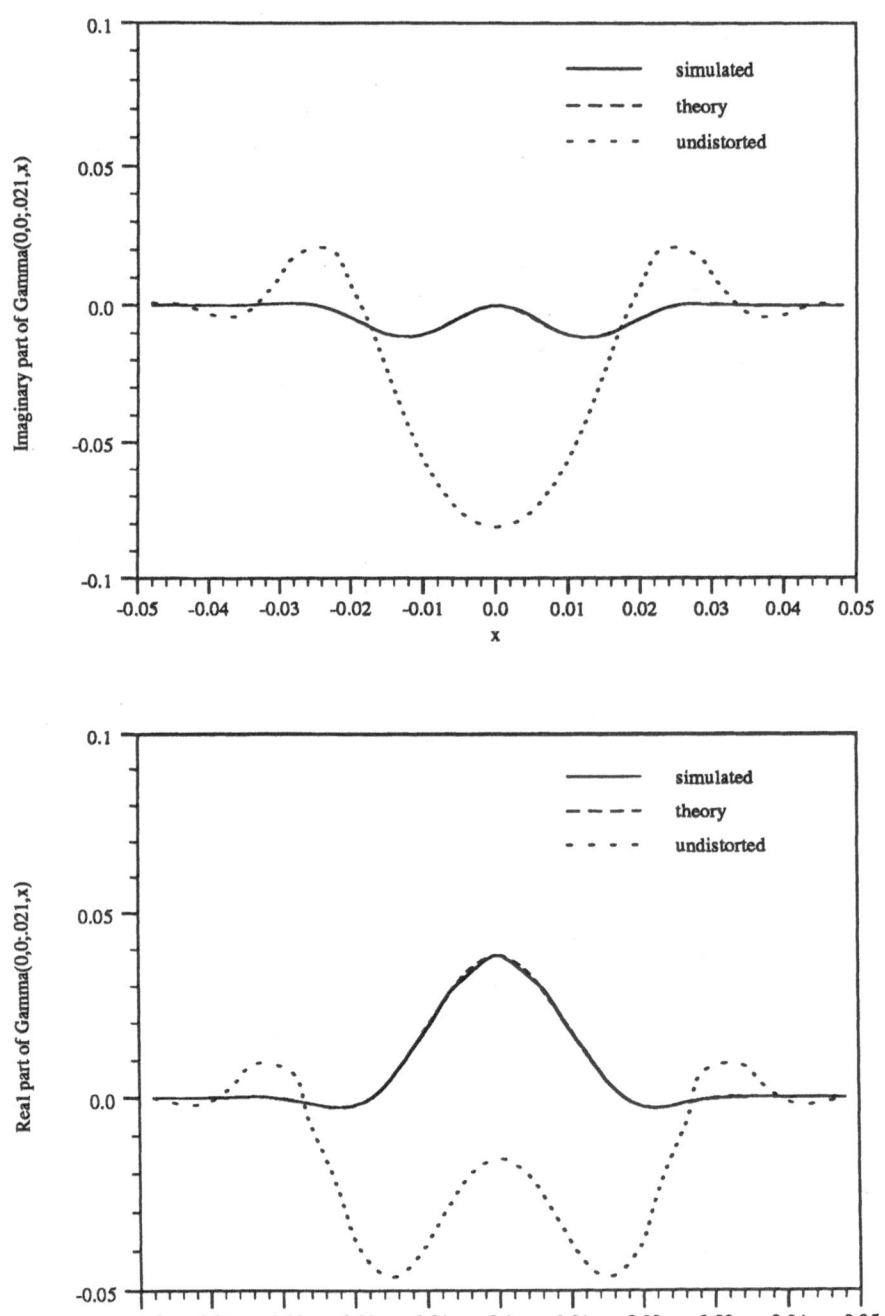

Figure 5.5: Average Coherence Function Off Boresight,$\Gamma(0,0;.021,x)$, At 1000 m, Based On 8000 Samples

Variable:	First Moment	Second Moment
I(0,0)	.12208	.02222
I(0,.021)	.03934	.00372
I(0,-.021)	.03924	.00372
I(.021,0)	.03840	.00359
I(-.021,0)	.03989	.00378
I(.021,.021)	.01416	.00068
I(.021,-.021)	.01412	.00069
I(-.021,.021)	.01475	.00079
I(-.021,.021)	.01465	.00076

Table 5.4: Moments of the Irradiance Function

V_t was smooth so we could consider its values as specific points. The complex valued random variable $V_t(\rho)$ is also Gaussian, and if its real and imaginary parts are considered to be independent and identically distributed, the magnitude $|V_t(\rho)|$ should obey a Rice-Nakagami distribution, which has the density

$$f(x) = \frac{x}{\sigma^2} e^{-\frac{x^2+A^2}{2\sigma^2}} I_0(\frac{Ax}{\sigma^2})$$

where A and σ^2 are the parameters of the distribution, and I_0 is the zeroth order Bessel function. If we assume that $\sqrt{I}$ has this distribution then I has moments

$$E(I) = 2\sigma^2 + A^2$$
$$E(I^2) = 8\sigma^4 + 8\sigma^2 A^2 + A^4$$

If the method of moments is used to estimate A and σ^2, the estimates are given by

$$\hat{A} = (2m_1^2 - m_2)^{1/4}$$
$$\hat{\sigma^2} = \frac{1}{2}(m_2 - \hat{A}^2)$$

where m_1 and m_2 are the sample first and second moments given in Table 5.4. For the results of the simulation with 8000 sample points, it was impossible to find positive A^2 and σ^2 to satisfy the above equations, hence the Rice-Nakagami distribution was rejected. This confirms the generally accepted view that the method of small perturbations is not valid for distances over 100 m, or in strong turbulence.

Under the method of smooth perturbations, or Rytov method, the log of $V_t(\rho)$ should be a Gaussian random variable, hence the log-normal distribution is considered for $I(\rho)$. In addition, some physicists have used the central limit theorem to intuitively justify this distribution, as the solution V can be seen as the limit of a product of random operators. It is actually much more difficult to apply the central limit theorem in this context, and the problem is more complex.

The log-normal distribution has the density

$$f(x) = \frac{1}{\sqrt{2\pi\sigma^2}x} e^{-(\ln(x)-m)^2/2\sigma^2}$$

Variable :	$\ln \bar{I}$	$Var(\ln I)$	$\bar{I}$	$pred.\bar{I}$	$Var(I)$	$pred.Var(I)$
$I(0,0)$	-2.367	.6669	.12208	.1309	.00730	.01624
$I(0,.021)$	-3.997	2.226	.03934	.05589	.00217	.02580
$I(0,-.021)$	-3.966	2.029	.03924	.05225	.00218	.01804
$I(.021,0)$	-4.003	2.059	.03840	.05110	.00212	.01786
$I(-.021,0)$	-3.954	2.086	.03989	.05440	.00219	.02087
$I(.021,.021)$	-5.220	2.532	.01416	.01918	.00048	.00426
$I(.021,-.021)$	-5.217	2.509	.01412	.01902	.00049	.00408
$I(-.021,.021)$	-5.215	2.605	.01475	.01999	.00058	.00501
$I(-.021,.021)$	-5.182	2.447	.01465	.01909	.00054	.00385

Table 5.5: Moments of the Irradiance Function, Simulated and Log-Normal Prediction

and the moments of I with this distribution are

$$E[I] = e^{m+\frac{1}{2}\sigma^2}$$
$$Var[I] = e^{2m+\sigma^2}(e^{\sigma^2}-1)$$

The maximum-likelihood estimates for m and σ^2 are

$$\hat{m} = \frac{1}{n}\sum_{j=1}^{n}\ln(I_j)$$
$$\hat{\sigma}^2 = \frac{1}{n}\sum_{j=1}^{n}(\ln(I_j)-\hat{m})^2$$

These estimates actually gave very poor fits for the distribution as can be seen from comparing the actual first two moments of I with the values predicted by the maximum likelihood estimates in Table 5.5.

Since the maximum likelihood estimates gave such poor results, the method of moments was used to estimate m and σ^2 from the first two sample moments of I. This hopefully removed some of the numerical difficulties of taking logarithms of very small numbers. The fits of the cumulative distributions corresponding to these estimates are given for three points in Figures 5.6 , 5.7 and 5.8. As can be seen from the graphs, this is not a very good representation of the simulated data.

The log-normal distribution was predicted by the method of smooth perturbations, or Rytov's method, which linearizes the equation for $\log(V)$. The failure of the simulated irradiance to fit a log-normal distribution is indicative of the breakdown of Rytov's method when it is applied to a distance of 1000 m. This also indicates that there is something wrong with using multiplicative central limit type arguments for this problem.

67

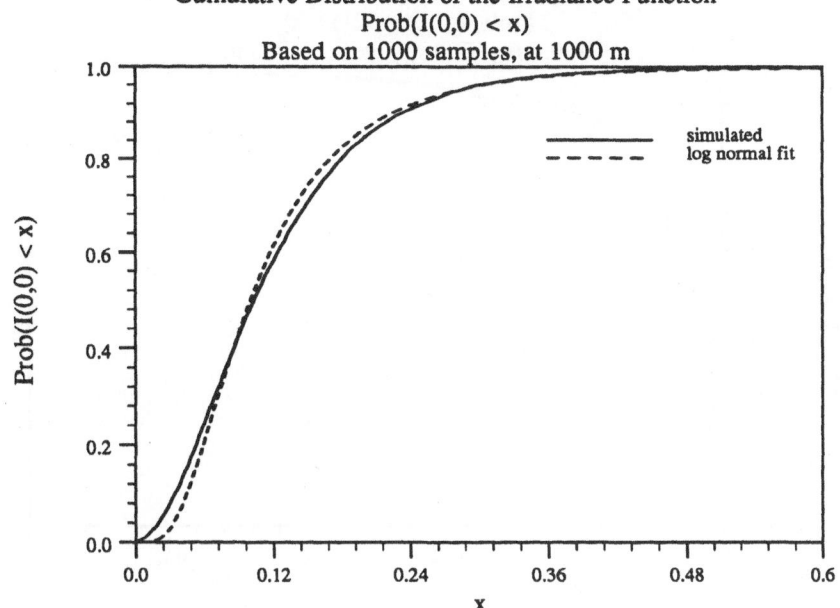

Figure 5.6: Log-Normal Fit to Simulated Irradiance Distribution for $I(0,0)$

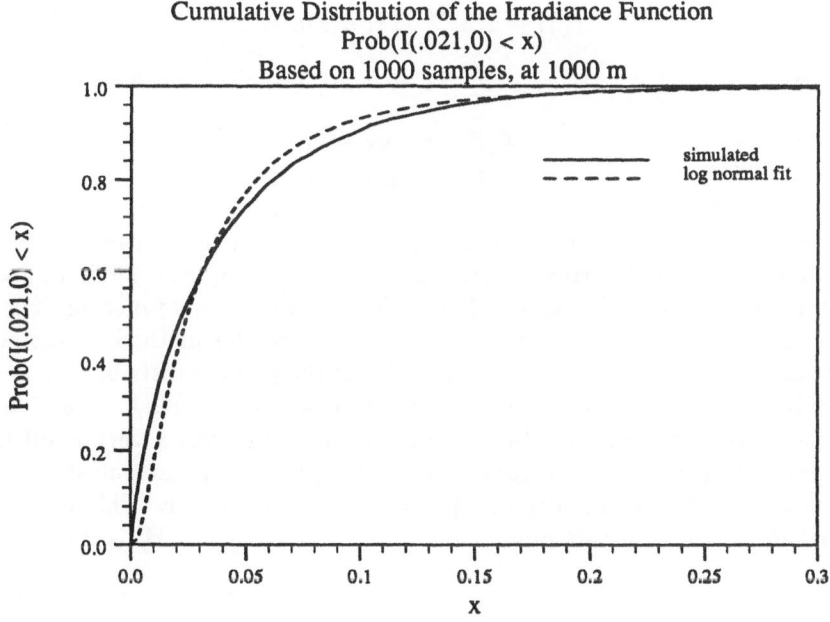

Figure 5.7: Log-Normal Fit to Simulated Irradiance Distribution for $I(.021,0)$

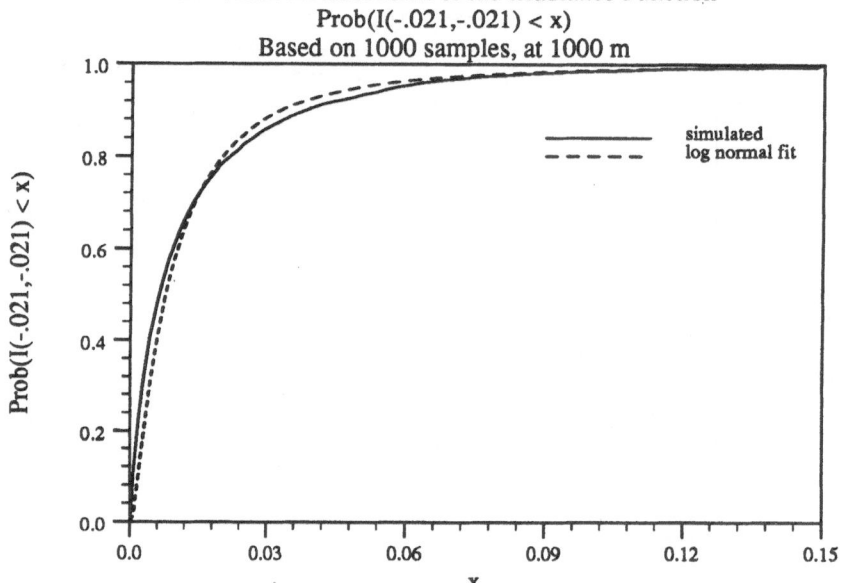

Figure 5.8: Log-Normal Fit to Simulated Irradiance Distribution for $I(-.021, -.021)$

Finally I consider the gamma, or Pearson type III, distribution, which has the density

$$f(x) = x^c e^{-x/d} / \Gamma(c+1) d^{c+1}$$

and moments

$$
\begin{aligned}
E[I] &= d(c+1) \\
Var(I) &= d^2(c+1)
\end{aligned}
$$

There is no theoretical basis for postulating a gamma distribution for the irradiance, however it did appear to describe the simulated irradiance distribution with good accuracy at all points where I was observed. Because of the Gamma function, it is difficult to use maximum likelihood to estimate c and d, hence the method of moments was employed, using the first two moments. The fits of the gamma distribution to the simulated irradiance distribution at three points are given in Figures 5.9 , 5.10 and 5.11. As can be seen from the graphs, the gamma distribution gives a very good fit at the boresight for $I(0,0)$, and a fairly good fit at $I(.021, 0)$. The fit at points more distant from the center of the beam, such as $I(-.021, -.021)$, is poorer. This was consistent with the results at the other six points, as can be seen in Figures B.1 - B.6 in Appendix B.

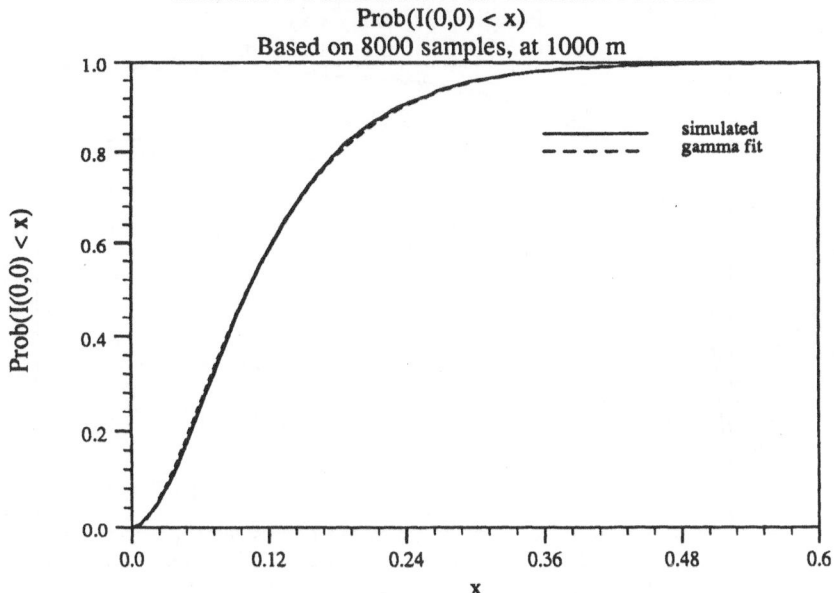

Figure 5.9: Gamma Fit to Simulated Irradiance Distribution for $I(0,0)$

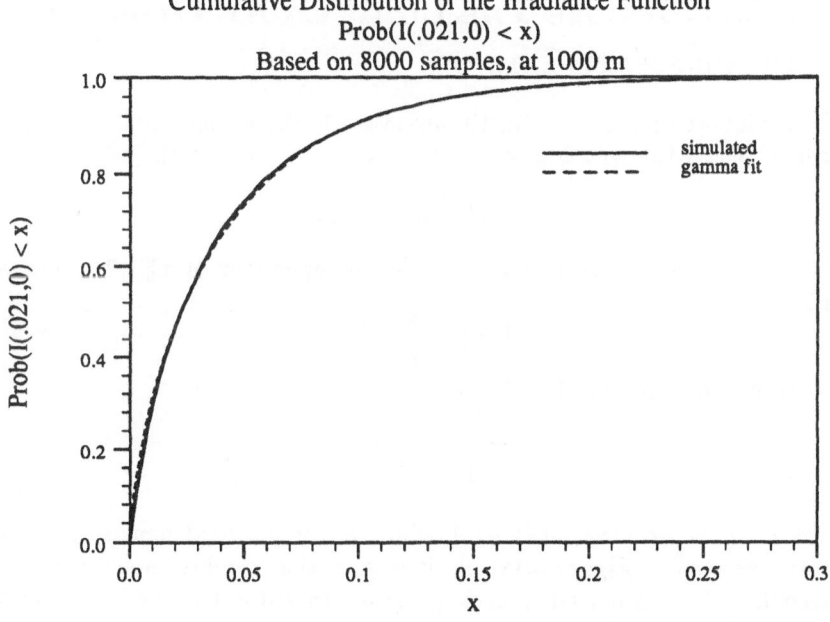

Figure 5.10: Gamma Fit to Simulated Irradiance Distribution for $I(.021,0)$

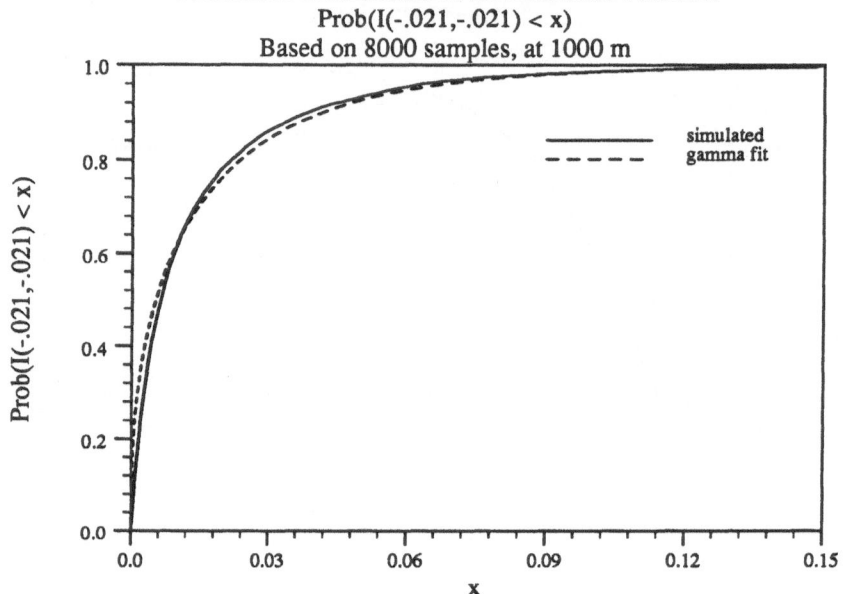

Figure 5.11: Gamma Fit to Simulated Irradiance Distribution for $I(-.021, -.021)$

5.7 White Noise As The Limit Of An Ornstein-Uhlenbeck Process: Theory

One of the simplest stationary stochastic processes is the Ornstein-Uhlenbeck process. It can be generated as the solution to

$$\dot{x}_t^a = -ax_t^a + aN_t, \quad x_0^a \text{given}$$

where N is a scalar valued white noise process independent of x_0^a. The steady state covariance is

$$R(\tau) = \frac{a}{2}e^{-a|\tau|}$$

and the corresponding spectral density is

$$\Phi(\lambda) = \frac{a^2}{a^2 + \lambda^2}$$

Note that as $a \to \infty$ we have $\Phi(\lambda) \to 1$ which is the spectral density corresponding to white noise, as seen in Figure 5.12. So it is seen that a white noise process can be approximated by an Ornstein-Uhlenbeck process. In Hilbert space terms $x^a = H^a N$ where H^a is a Hilbert-Schmidt operator converging strongly to the identity as $a \to \infty$. Recall that in the definition of a physical random variable, white noise was approximated by finite dimensional projections $P_n N \to N$ as $n \to \infty$.

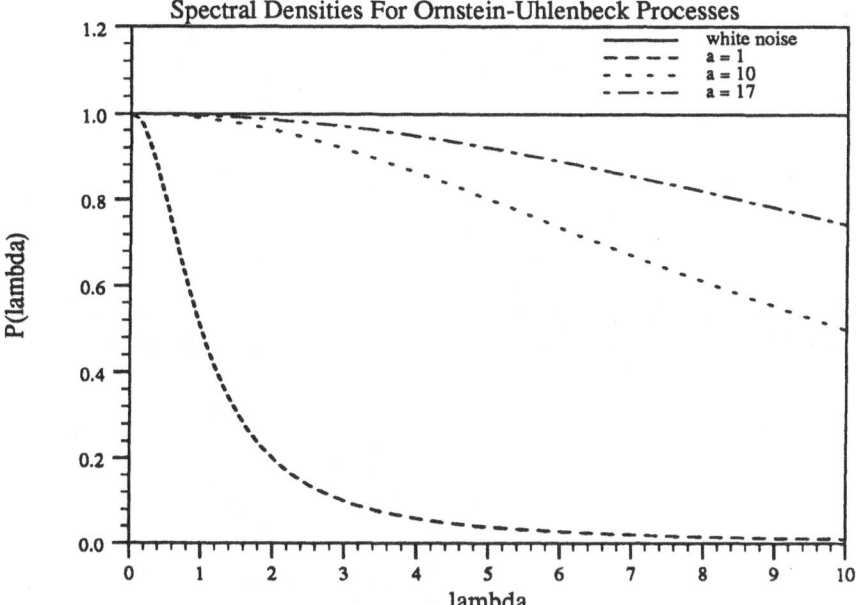

Figure 5.12: Spectral Densities For Ornstein-Uhlenbeck Processes With Parameter a

Hence instead of the white noise process used in the Markov approximation, we use an n_1 process with the covariance function

$$E[n_1(t, \rho)n_1(t', \rho')] = e^{-a|t-t'|}\frac{a}{2}A(\rho - \rho')$$

and observe what happens as $a \to \infty$. The results of this simulation will yield evidence of the appropriateness or inappropriatness of the Markov approximation.

In terms of the abstract bilinear system model, the system is represented by

$$\dot{V}_t = \frac{i}{2k}\nabla^2 V_t + B(V_t, N_t^a) , \ V_0 \in H \qquad (5.7)$$

$$\dot{N}_t^a = -aN_t^a + aN_t , \ N_0^a \in H_r \qquad (5.8)$$

where N is our usual white noise and N_0 is an H_r valued white noise independent of N with covariance $\frac{a}{2}I$.

To see that this is a valid approximation method, I first show that $N^a \to N$ strongly which implies that $V(N^a) \to V(N)$ strongly both in $\mathcal{H}$ and in H for each t.

Proposition 5.7.1 *Suppose $N \in \mathcal{W} = L_2[(0, T); H_r]$, H_r a real separable Hilbert space, and $N_0 \in H_r$. Then N^a defined in Equation 5.8 converges strongly to N in $\mathcal{W}$ as $a \to \infty$.*

Proof
N^a can be written as

$$N^a = N^{0,a} + H^a N$$

where

$$N_t^{0,a} = e^{-at}N_0^a$$

$$(H^a N)_t = \int_0^t ae^{-a(t-s)}N_s \, ds$$

In order to show N^a converges to N I show $N^{0,a} \to 0$ and $H^a N \to N$. First,

$$\|N^{0,a}\|^2 = \frac{1}{2a}(1 - e^{-2aT})\|N_0^a\|^2$$

which goes to zero as a goes to infinity.

Now consider $N \in \mathcal{W}$ such that N has an L_2 derivative in $\mathcal{W}$. Then,

$$\|N_t - \int_0^t ae^{-a(t-s)}N_s \, ds\| = \|e^{-at}N_0 + \int_0^t e^{-a(t-s)}N_s' \, ds\|$$

$$\le e^{-at}\|N_0\| + \frac{1}{\sqrt{2a}}\|N'\|$$

Hence

$$\|N - H^a N\| \le \frac{1}{\sqrt{2a}}(\|N_0\| + T\|N'\|)$$

Hence $H^a N \to N$ strongly for absolutely continuous N. Since the absolutely continuous functions are dense in $\mathcal{W}$ it is enough to show that the H^a are uniformly bounded to show that $H^a N$ converges for all N in $\mathcal{W}$. Now

$$\|H^a N\|^2 = \int_0^T \|\int_0^t ae^{-a(t-s)}N_s \, ds\|^2 \, dt$$

$$= \int_0^T \int_0^t \int_0^t a^2 e^{-a(t-s+t-r)}[N_s, N_r] \, dsdrdt$$

$$= \int_0^T \int_0^T \frac{a}{2}(e^{-a|s-r|} - e^{-a(2T-s-r)})[N_s, N_r]$$

$$= \int_0^T \int_0^s ae^{-a(s-r)}[N_s, N_r] \, drds - \frac{a}{2}\|\int_0^T e^{-a(T-s)}N_s \, ds\|^2$$

$$\le [N, H^a N]$$

$$\le \|N\|\|H^a N\|$$

Hence, $\|H^a\| \le 1$ for all a. Hence $H^a N \to N$ strongly in $\mathcal{W}$. $\square$

It is difficult to show that the solution corresponding to N^a converges in mean square, hence I consider a product form solution $V^{l,a}$. Matters are also simplified by taking a from the discrete set of integers. Finally, note that the spectral densities for $n_1^a = w * N^a$ are monotonically increasing in a, i.e. for

$$\Phi^a(\mu, \lambda) = \frac{a^2}{a^2 + \mu^2}\Phi(\lambda)$$

we have

$$\Phi^a(\mu, \lambda) \le \Phi^b(\mu, \lambda), \; b \ge a$$

In terms of the covariance operator, this implies that

$$R^a(\tau, \rho) = \frac{a}{2}e^{-a|\tau|}A(\rho)$$

we have

$$R^a \leq R^b, \ b \geq a$$

as an operator on $L_2(\mathbf{R}^3)$. Hence it is possible to generate a sequence of homogeneous random fields n_1^a such that $n_1^b - n_1^a$ is independent of n_1^a for all $a < b$. This property will be used to prove that for n_1^a chosen in this fashion the random variables $V'(N^a)$ are Cauchy in mean square. First, an important lemma is needed.

Lemma 5.7.2 *Let*

$$N_t^M = \int_0^T M(t,s)N_s \ ds$$

where M is square integrable on $[0,T]^2$, that is M is the kernel of a Hilbert-Schmidt operator. Then the product forms $V^l(N^M)$ and $V_t^l(N^M)$ are physical random variables as well.

Proof
The proof is identical to that for Theorem 4.4.1 with the exception that in Equation 4.6 $g^{n,m}$ is now

$$g_\rho^{n,m} = 1 -$$
$$\exp\{\frac{-k^2}{2}\sum_{p=m+1} n(\int_{jT/l}^{2t-jT/l}\int_0^T M(t,s)\int_{\mathbf{R}^2} w(\rho-\rho')e_p(s,\rho') \ d\rho \ ds \ dt)^2\}$$

And since M has a square integrable kernel, the rest of the proof proceeds as in Theorem 4.4.1. □
At this point the following Theorem can be proven.

Theorem 5.7.3 *Let N^a be a sequence of weak Gaussian random variables in $\mathcal{W}$ with zero mean and covariance*

$$E([x, N^a(t)][y, N^a(s)]) = \frac{a}{2}e^{-a|t-s|}[x,y]$$

where $x, y \in H_r$. Denote

$$R^a(t) = \frac{a}{2}e^{-a|t|}$$

It is possible to represent N^a as

$$N_t^a = \int_0^T \sum_{j=1}^a M^j(t,s)N_s \ dsd\rho' \tag{5.9}$$

where the kernels M^j are square integrable and

$$\int_0^T M^j(t,s)M^k(\tau,s) \ ds = R^j(t-\tau)R^{j-1}(t-\tau) \tag{5.10}$$

when $k = j$ and zero otherwise. Let $V^{l,a}$ be the product form solution

$$V_t^{l,a} = \Theta_t^{l,a}[\prod_{j=1}^{j=\lfloor tl/T\rfloor} e^{ik\int_{(j-1)T/l}^{jT/l} L(N^a(\tau))\ d\tau} S_{T/l}]V_0$$

where Θ_t^l is as in Equation 4.5, only now with $L(N^a)$ instead of $L(N)$. That is

$$V_t^{l,a} = \Theta_t^{l,a} V_{\lfloor tl/T\rfloor}^{l,a}$$

Then $V_t^{l,a}$ is Cauchy in mean square and so is $V^{l,a}$

Proof
First the representation for N^a must be verified. Let N^a be represented as in Equation 5.9. Since $R^{j+1} \geq R^j$ it is possible to choose M^j satisfying Equation 5.10 by taking $M^j(t,s)$ to have support on $[T(1-2^{-(j-1)}), T(1-2^{-j})]$. Hence it is possible to represent N^a in this manner.

By Lemma 5.7.2 the product form $V^{l,a}$ is a physical random variable for all a and it is possible to talk about its moments.

Denote by t_j the points $t_j = jT/l$. Also, denote

$$N_j^a = \int_{(j-1)T/l}^{jT/l} N^a(\tau)\ d\tau$$

$$\|V_t^{l,b} - V_t^{l,a}\| \leq \sum_{j=0}^{j=\lfloor tl/T\rfloor} \|(e^{L(N_j^b)} - e^{L(N_j^a)})S_{T/l}V_{t_j}^{l,a}\|$$
$$+ \|(\Theta_t^{l,b} - \Theta_t^{l,a})V_{\lfloor tl/T\rfloor}^{l,a}\|$$

Clearly it is sufficient to show that each term in the sum above can be made arbitrarily small by choosing a, $b > a_0$. I consider one term of the sum.

$$E\|(e^{L(N_j^b)} - e^{L(N_j^a)})S_{T/l}V_{t_{j-1}}^{l,a}\|^2 = \tag{5.11}$$
$$= E[(2 - e^{L(N_j^b - N_j^a)} - e^{L(N_j^a - N_j^b)})S_{T/l}V_{t_{j-1}}^{l,a}, S_{Tl}V_{t_{j-1}}^{l,a}] \tag{5.12}$$
$$= E[(2 - 2e^{\frac{1}{2}(r_l^b - r_l^a)D})S_{T/l}V_{t_{j-1}}^{l,a}, S_{T/l}V_{t_{j-1}}^{l,a}] \tag{5.13}$$

using the Fubini theorem, where

$$D = \sum_{k=1}^{\infty} L^2(\phi_k)$$

for $\{\phi_k\}$ CON in H_r and

$$r_l^a[x,y] = E\int_0^{T/l}\int_0^{T/l} [N_s^a, x][N_t^a, y]\ dsdt$$
$$= \int_0^{T/l}\int_0^{T/l} R^a(t-s)\ dtds\ [x,y]$$
$$= [T/l - (1 - e-aT/l)/a][x,y]$$

In this case, $D = -k^2\|w\|^2$ Now,

$$
\begin{aligned}
r_l^b - r_l^a &= (1 - e^{-aT/l})/a - (1 - e^{-bT/l})/b \\
&\leq 1/a
\end{aligned}
$$

for $a \leq b$. Now the term in Equation 5.13 is equal to

$$
2(1 - e^{-k^2\|w\|^2(r_l^b - r_l^a)})\|V_0\|^2
$$

which is smaller than

$$
2(1 - e^{-k^2\|w\|^2/a})\|V_0\|^2
$$

and this quantity goes to zero as $a \to \infty$. A similar argument can be used for the Θ term. Hence each term is Cauchy in mean square, hence $V_t^{l,a}$ is Cauchy in mean square. Since $V_t^{l,a}$ is uniformly bounded in a and t this is enough to imply that $V^{l,a}$ is Cauchy in mean square as well. $\square$

Using the representation from the Theorem 5.7.3, it was shown that $V^{l,a}$ was Cauchy in mean square. Because of the representation used, it is not true that $V^{l,a}$ converges to V^l in mean square. In the following Theorem, it is shown that if N^a is generated as an Ornstein-Uhlenbeck process run 'backward', then $V_t^{l,a} \to V_t^l$ in mean square.

Theorem 5.7.4 *Let N^a be the $\mathcal{W}$ valued weak random variable represented as*

$$
N^a = \int_t^T a e^{-a(s-t)} N_s \, ds \tag{5.14}
$$

where N is a $\mathcal{W}$ valued white noise. Then the product forms $V_t^{l,a}$ converge to V_t^l in mean square. In addition $V^{l,a}$ converges to V^l as an $\mathcal{H}$ valued random variable.

Proof

We wish to show that the product form

$$
V_t^{l,a} = \Theta_t^{l,a}[\prod_{j=1}^{j=\lfloor tl/T \rfloor} e^{L(N_j^a)} S_{T/l}]V_0
$$

where

$$
N_j^a = \int_{t_{j-1}}^{t_j} N_s^a \, ds
$$

converges in mean square to

$$
V_t^l = \Theta_t^l[\prod_{j=1}^{j=\lfloor tl/T \rfloor} e^{L(N_j)} S_{T/l}]V_0
$$

where

$$
N_j = \int_{t_{j-1}}^{t_j} N_s \, ds
$$

and $t_j = jT/l$. As in the proof of Theorem 5.7.3,

$$
\begin{aligned}
\|V_t^l - V_t^{l,a}\| &\leq \sum_{j=0}^{j=\lfloor tl/T \rfloor} \|(e^{L(N_j - N_j^a)} - 1)S_{T/l}V_{t_{j-1}}^l\| \\
&+ \|(\Theta_t^l - \Theta_t^{l,a})V_{\lfloor tl/T \rfloor}^{l,a}\|
\end{aligned}
$$

and it is sufficient to show that each term converges to zero.

First, it is helpful to note that

$$N_j - N_j^a = (N^a(t_j) - N^a(t_{j-1}))/a$$

Hence for each $x \in H_r$

$$E[(N_j - N_j^a, x)^2] \leq \frac{2}{a}\|x\|^2$$

Also, because of the way in which N^a was defined, $N_j^a - N_j$ is uncorrelated with N_k for $k < j$. Hence,

$$
\begin{aligned}
E\|(e^{L(N_j^a - N_j)} - I)S_{T/l}V_{t_{j-1}}^l\|^2 &= \\
&= E[(2 - e^{L(N_j^a - N_j)} - e^{L(N_j^a - N_j)})S_{T/l}V_{t_{j-1}}^l, S_{T/l}V_{t_{j-1}}^l] \\
&= E[(2 - 2e^{\frac{1}{2}d_j^a D}S_{T/l}V_{t_{j-1}}^l, S_{T/l}V_{t_{j-1}}^l]
\end{aligned}
$$

using the Fubini theorem and the independence of $N_j^a - N_j$ and N_k for $k < j$, where

$$d_j^a = E(N_j - N_j^a, x)^2/\|x\|^2 \leq 2/a$$

Hence,

$$E\|(e^{L(N_j^a - N_j)} - I)S_{T/l}V_{t_{j-1}}^l\|^2 \leq 2(1 - e^{-k^2\|w\|^2/a})\|V_0\|^2$$

and this quantity converges to zero as a goes to infinity. Since $V_t^{l,a} - V_t^l$ is uniformly bounded in l and t, this is enough to imply that $V^{l,a}$ converges to V^l in mean square as a $\mathcal{H}$ valued random vector. $\square$

5.8 White Noise As The Limit Of An Ornstein-Uhlenbeck Process: Simulation

In this section, I compare the coherence function and distribution of the irradiance corresponding to a white noise input to those for Ornstein-Uhlenbeck processes with parameter $a = .5$, $a = .02$ and $a = .01$. A Monte Carlo simulation useing 4000 runs for each of the above values of a was performed. Monte Carlo simulations using 1000 runs were also performed for Ornstein-Uhlenbeck processes with $a = .1, 2, 20$. From the last section, we expect that as $a \to \infty$ the coherence function and irradiance function will converge as well. The simulation gives us an idea of what happens for finite a.

For digital simulation, it is necessary to discretize the Ornstein-Uhlenbeck process, and this is done by integrating it over disjoint intervals. For example, if x^a is a scalar valued Ornstein-Uhlenbeck process, define the sequence

$$x_k^a = \int_{(k-1)L}^{kL} x_t^a \, dt$$

Then the covariance for the discretized sequence is

$$
\begin{aligned}
E[x_k^a x_k^a] &= L - (1 - e^{-aL})/a \\
E[x_k^a x_{k+p}^a] &= e^{-aL(p-1)}(1 - e^{-aL})^2/2a \ , \ p > 0
\end{aligned}
$$

	$a = .01$	$a = .1$	$a = .5$	$a = 1$	$a = \infty$
R_0	.123	1.07	3.16	4.01	5
R_1	.119	.774	.843	.493	0
R_2	.113	.470	.0692	.0033	0
R_3	.108	.285	.0057	.00002	0

Table 5.6: Covariance For Discretized Ornstein-Uhlenbeck Process, $L = 5$

	$a = .01$	$a = .1$	$a = .5$	$a = 1$	$a = \infty$
R_0/L	.9	.99	.998	.999	1
R_1/L	.05	.005	.001	.0005	0
R_2/L	2×10^{-6}	≈ 0	≈ 0	≈ 0	0

Table 5.7: Normalized Covariance For Ornstein-Uhlenbeck Process, $L = 1000$

A stochastic sequence with this covariance can be represented and generated using a first order ARMA model, as is done in the simulation. Values of the covariance R_p^a for x_k^a are given in Table 5.6 for the case where $L = 5$, that is the step size for the simulation.

Although I expected the results for $a = .1$, $a = .5$ and $a = 1$ to differ significantly from the white noise case, $a = \infty$, this was not the case. This appeared to be due to the fact that the as the beam propagates, the turbulence field is approximately integrated again, hence the variables of interest are R_p/L for the case where $L = 1000$, the entire propagation length. The values are shown in Table 5.7.

From Table 5.7, it is easy to see why the simulated results resemble those for a white noise input except when a is very small. This gives very strong confirmation of the accuracy of the Markov approximation, at least for long propagation distances.

If the irradiance function $I_t(\rho)$ is averaged over the simulation runs, the result describes the average spatial distribution (in x and y) of energy in the beam crossection at a distance t from the initial point.

The average irradiance for white noise inputs is virtually identical to that for an Ornstein Uhlenbeck process with parameter $a = .5$ or $a = .1$. A graphical comparison of the average irradiance for the white noise case and an O-U process with $a = .5$ is given in Figure 5.13 . $a = .5$ corresponds to a correlation length of about 2 m. Recall from Figure 5.12 that the spectral density for $a = .5$ does not closely resemble that for a white noise. Yet the results of the simulation are almost identical!

Figure 5.14 compares the average irradiance for a white noise input, and for Ornstein-Uhlenbeck processes with parameters $a = .01$ and $a = .02$. The white noise input corresponds to the case where $a = \infty$. It is easy to see that as the parameter a is decreased, the average irradiance across the boresight experiences less spreading.

The shape of the average irradiance across the boresight is very well described by a

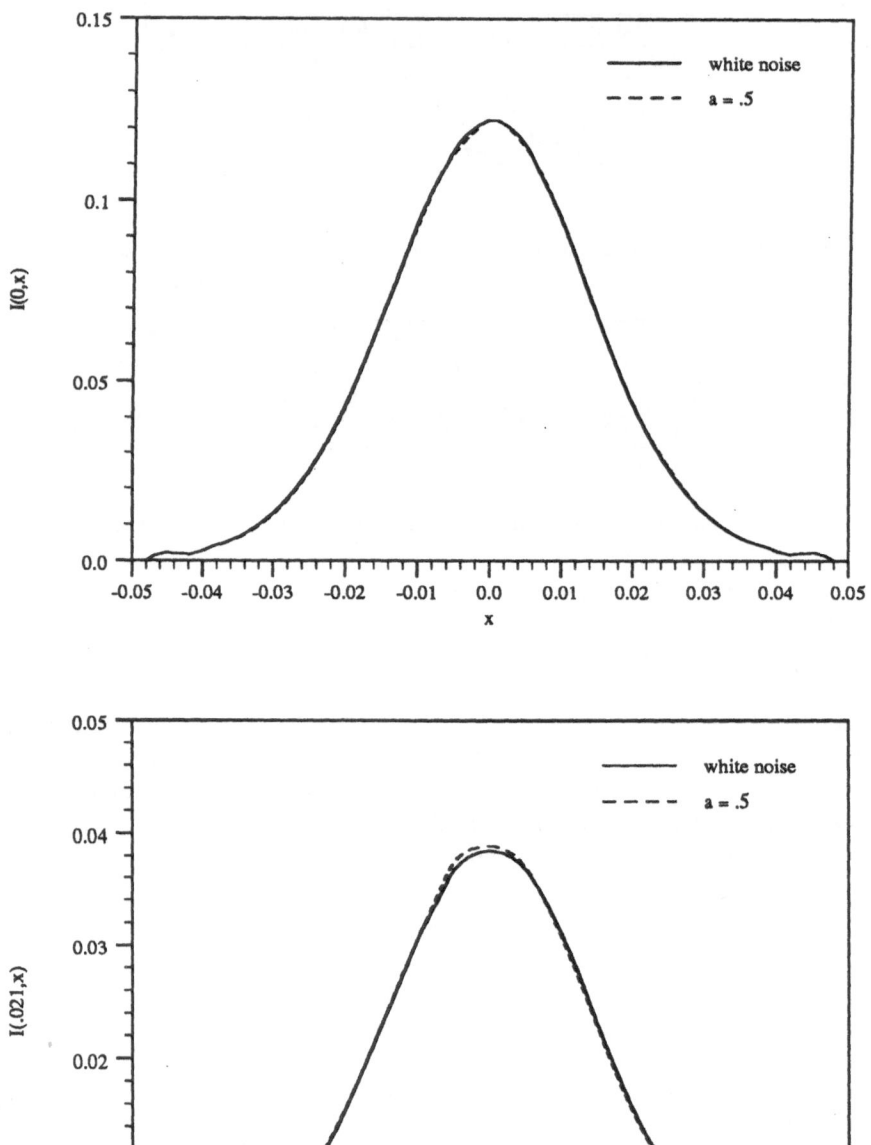

Figure 5.13: Average Irradiance for White Noise and O-U Input $a = .5$, Across Bore-sight, $I(0,x)$, And Off Boresight, $I(.021,x)$, Based On 8000, 4000 And 4000 Samples, At 1000 m

Gaussian curve of the form

$$f(x) = pe^{-x^2/2\sigma^2}$$

where p is the peak value and σ describes the spreading. The values of p and σ for the average irradiance $I(0, x)$ are given in Table 5.8

	p	σ
white noise	.122	.01421
$a = .02$	.129	.01387
$a = .01$	.135	.01357

Table 5.8: Gaussian Parametrization of Average Irradiance $I(0, x)$

The sample averages of the coherence function also show rapid convergence as the parameter a is increased. The correlation of V at the boresight with values along lines passing through the boresight and displaced from it by .021 m, $\Gamma(0, 0; 0, x)$ and $\Gamma(0, 0; .021, x)$, are considered. The average coherence for the case where $a = .5$ is once again identical to that for the white noise case compared to that for white noise case, as can be seen in Figures 5.15 and 5.16. Again, it is evident that the results are identical, and the solution for the Ornstein-Uhlenbeck process with $a = .5$ behaves like that for the white noise input in this regard.

It is more interesting to compare the results for the average coherence for an O-U process with $a = .01$ or $a = .02$ with that for white noise. The sample average coherence for these values of a is compared with that for white noise in Figures 5.17 and 5.18. When looking at the coherence across the boresight, the effect of increasing a appears to mostly be an attenuation of the peaks for both the real and imaginary parts. From Figure 5.4 it was apparent that the effect of the turbulence was both to attenuate the peaks of the coherence function and shift them some.

For the coherence of the beam at the boresight with values on a line displaced by .021 from the boresight, very little change is seen when a is increased from .01 to ∞. Again, the only effect is a slight attenuation of the peaks as a increases.

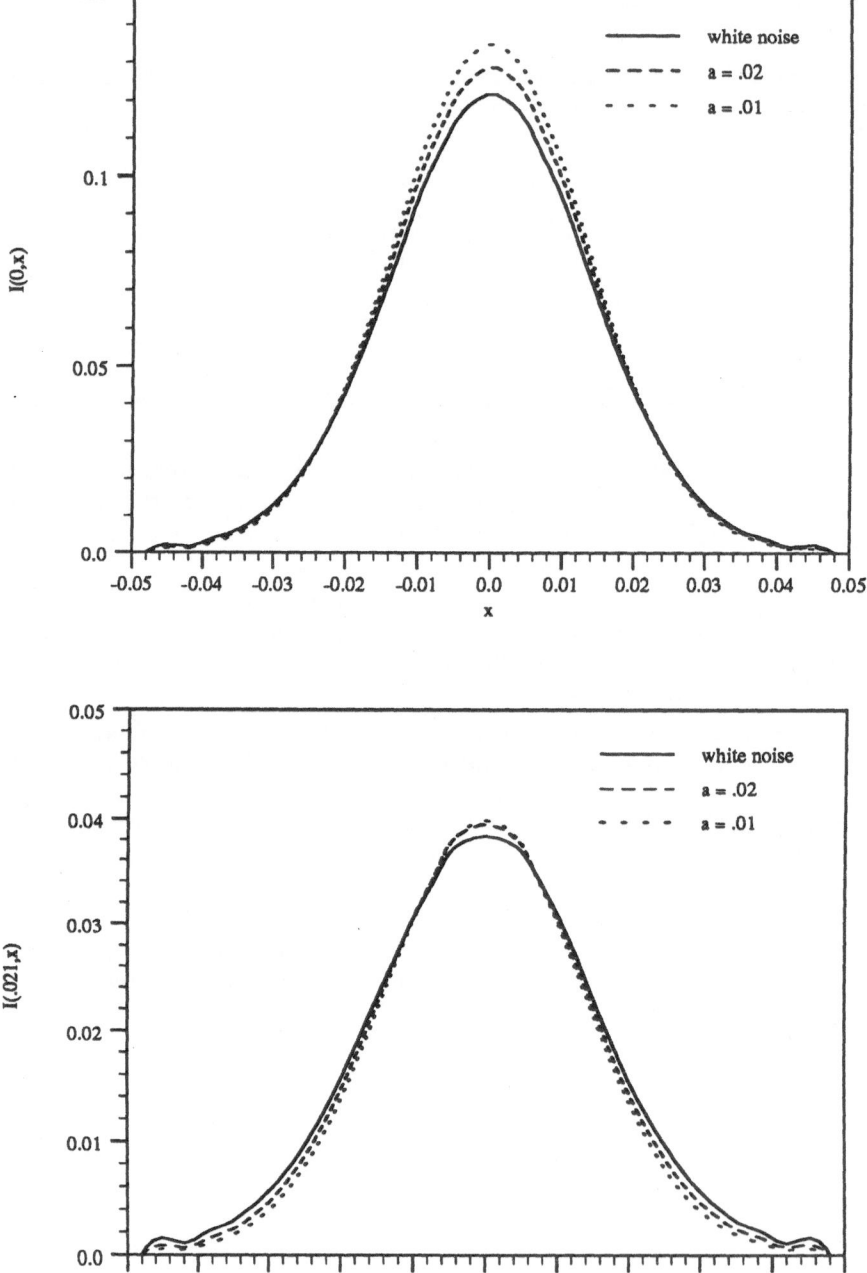

Figure 5.14: Average Irradiance for White Noise and O-U Input $a = .01$, $.02$, Across Boresight, $I(0, x)$, And Off Boresight, $I(.021, x)$, Based On 8000, 4000 And 4000 Samples, At 1000 m

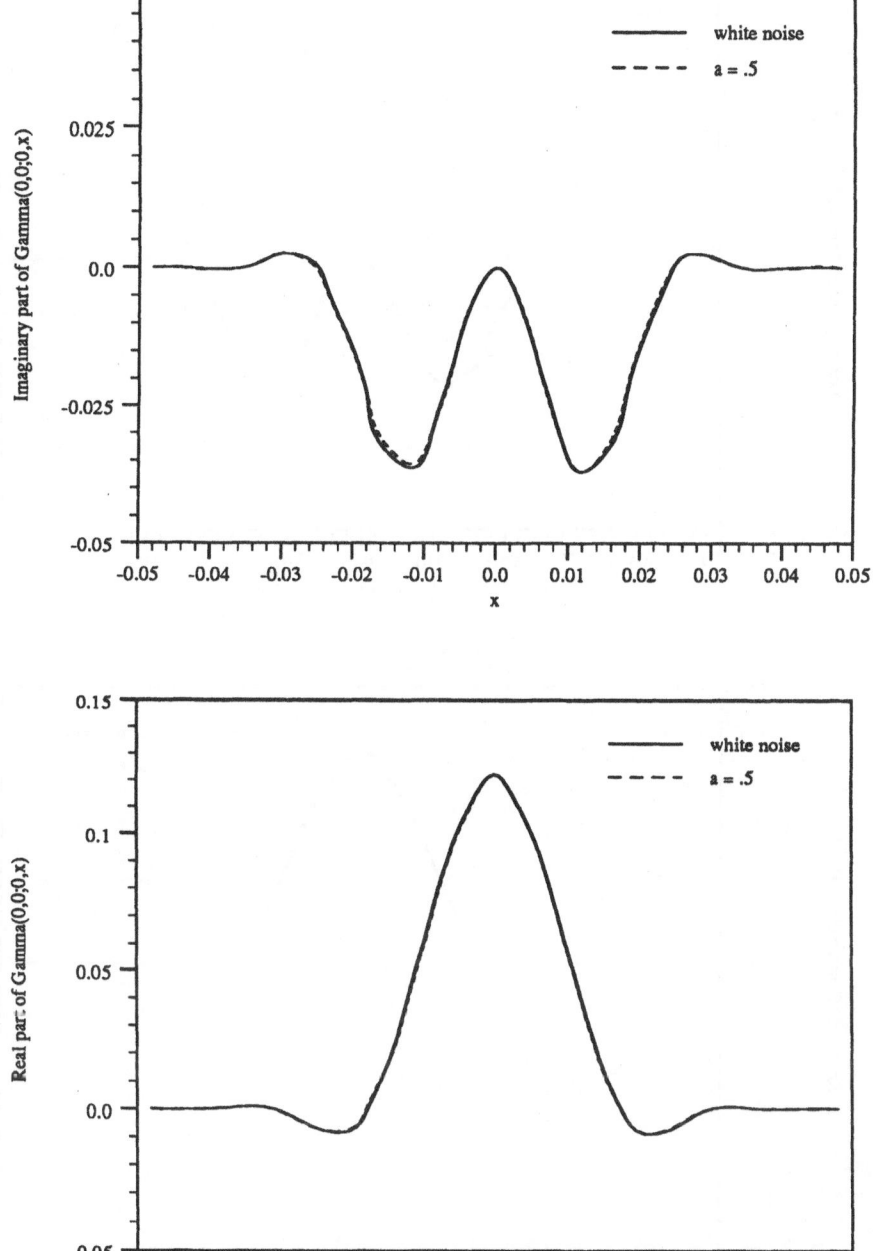

Figure 5.15: Average Coherence Function Across Boresight, $\Gamma(0,0;0,x)$ For White Noise And O-U Input $a = .5$, Based On 8000 and 4000 Samples, At 1000 m

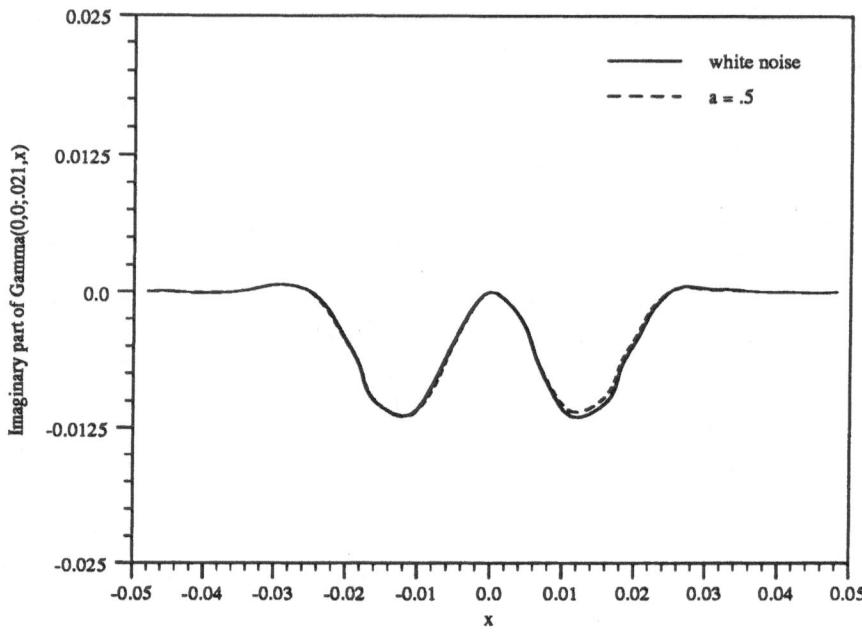

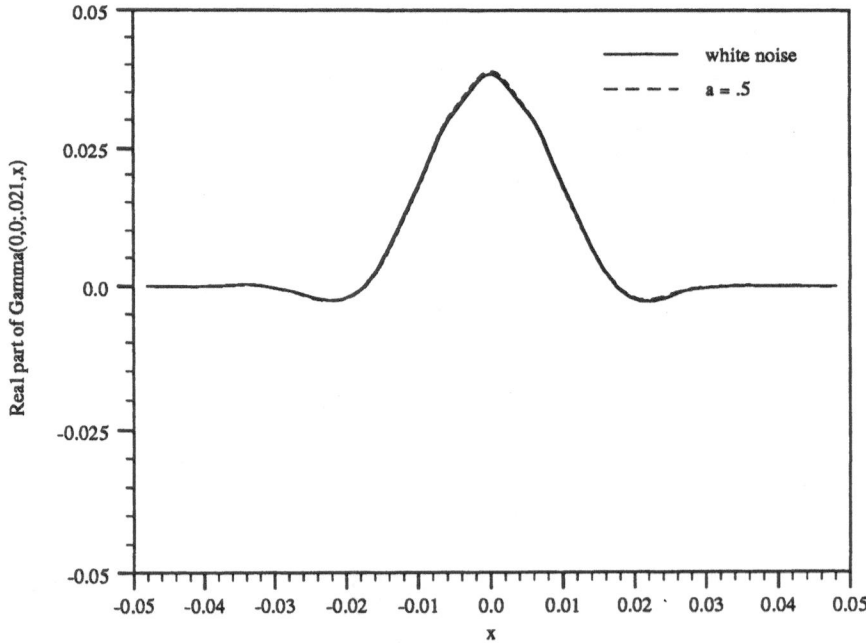

Figure 5.16: Average Coherence Function Off Boresight $\Gamma(0,0;.021,x)$ For White Noise And O-U Input $a = .5$, Based on 8000 and 4000 Samples, At 1000 m

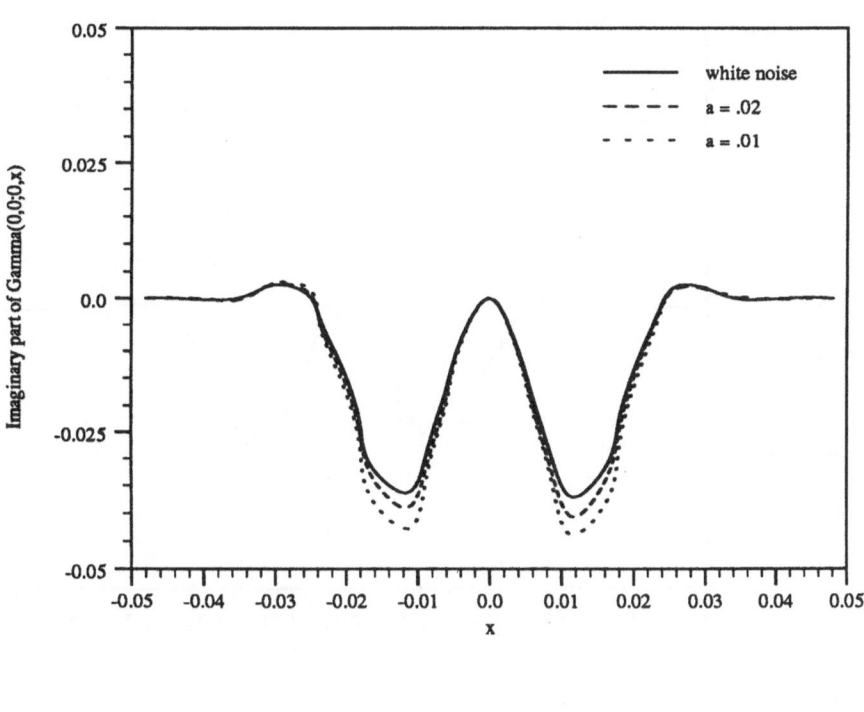

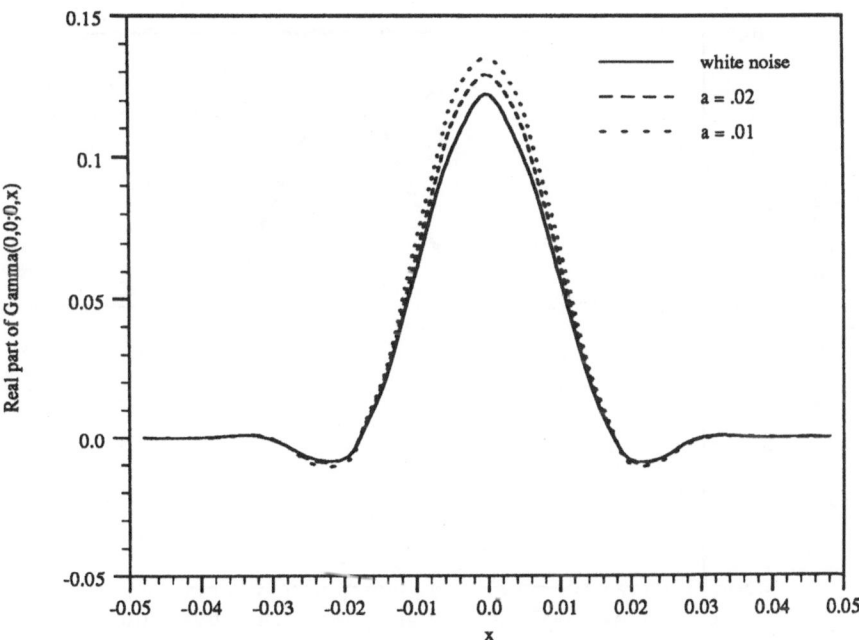

Figure 5.17: Average Coherence Function Across Boresight, $\Gamma(0,0;0,x)$ For White Noise And O-U Input $a = .01, .02$, Based On 8000, 4000 And 4000 Samples, At 1000 m

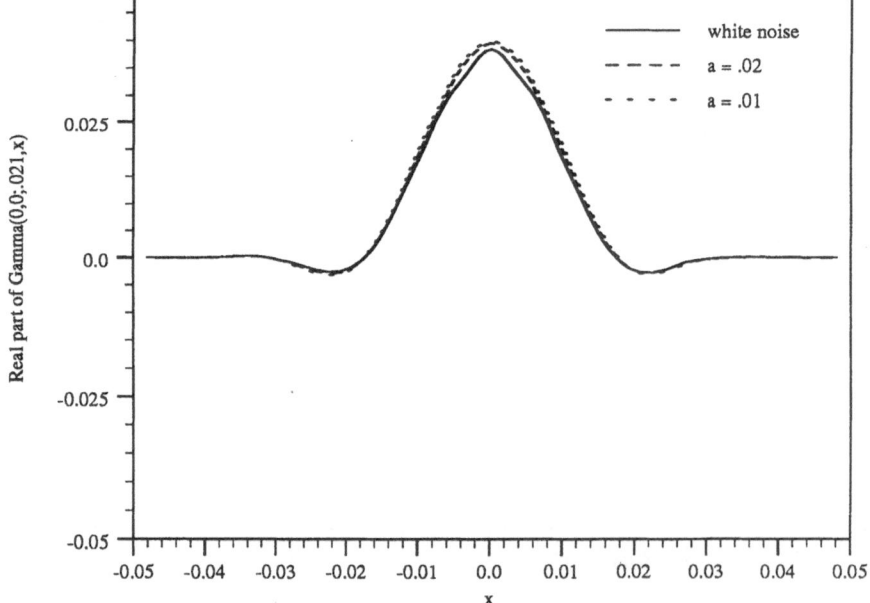

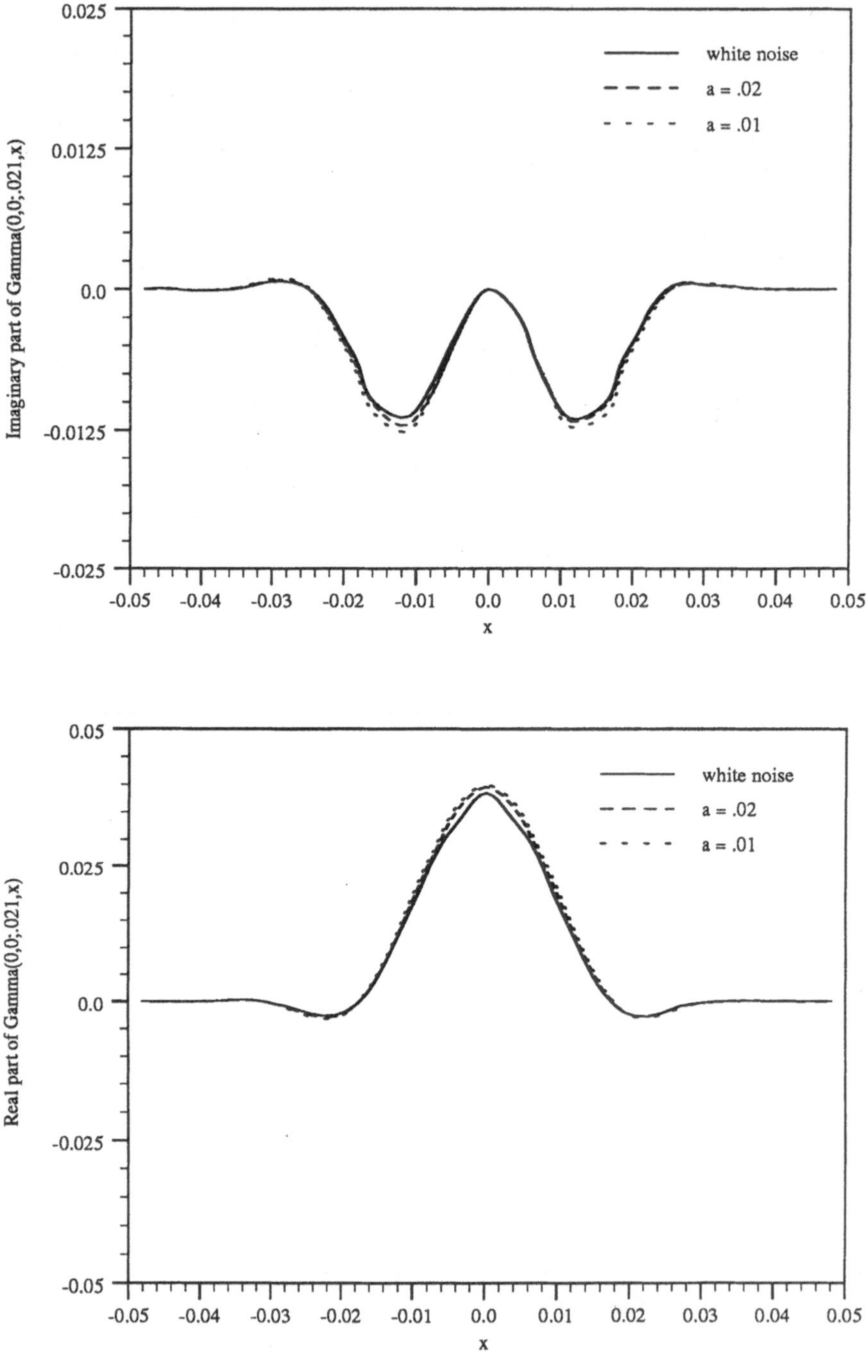

Figure 5.18: Average Coherence Function Off Boresight $\Gamma(0,0;.021,x)$ For White Noise And O-U Input $a = .01, .02$, Based on 8000, 4000 and 4000 Samples, At 1000 m

Finally, I consider the behavior of the probability distribution of the irradiance as the parameter a is increased. Once again, very rapid convergence is seen, and the distributions for $a = .5$ are indistinguishable from those for the white noise case, as seen in Figures 5.19 - 5.21.

Comparisons of the cumulative distributions of the irradiance for $a = .01$, $a = .02$ and white noise are given in Figures 5.22 - 5.24. Here some more interesting, although slight, effects can be seen. In Figure 5.22 it can be seen that the effect of increasing the bandwidth (increasing a) on the distribution of $I(0,0)$ is to raise the distribution curve, which corresponds in general to decreasing the irradiance at $(0,0)$, or shifting the density to the left.

In Figure 5.23 it can be seen that changing a has very little effect on the distribution of $I(.021,0)$. However, it is not true that increasing a has less effect further from the boresight, as can be seen in Figure 5.24 in the distribution of $I(-.021,-.021)$. At this point the effect of increasing a (increasing the bandwidth) is to lower the distribution curve, which corresponds to increasing $I(-.021,-.021)$ in general, and a shifting of the density to the right. This could be due either to increased spreading or increased bending of the beam, or both.

The means, standard deviations and skewness of $I(0,0)$ for $a = .5$, .02, .01 and white noise input are given in Table 5.9. The skewness parameter describes the lack of symetry of a probability distribution, and is defined as

$$a_3 = (E[(I - E(I))^3])^{1/3}$$

In addition the parameters c and d for the fitted gamma distribution are also shown.

	mean	standard deviation	skewness	c	d
white noise	.1221	.0855	.1663	1.037	.05994
a = .5	.1228	.08575	.1656	1.034	.06013
a = .02	.1291	.08688	.1739	1.208	.05846
a = .01	.1353	.08715	.1802	1.409	.05615

Table 5.9: Comparison Of $I(0,0)$ Distribution Parameters For White Noise And $a = .5$, .02, .01

The mean appears to decrease as a is increased, indicating more spreading or bending of the beam. The standard deviation decreases slightly as a is increased, which is not entirely expected. If a white noise input represents the maximum random disturbance, intuitively the standard deviation of the output should increase as the input approaches a white noise.

The skewness parameter is always positive here, indicating that the distribution is skewed to the right. It becomes less so as the bandwidth is expanded (a increases).

The parameter d for the gamma distribution is more or less a scale parameter. d increases as the bandwidth is expanded (a increases), which indicates a spreading of the distribution of $I(0,0)$. The parameter c for the gamma distribution is a shaping parameter. For $-1 < c < 0$ the density is unimodal with an infinite spike at the origin.

For $c > 0$ the density is unimodal with a peak somewhere to the right of the origin. The mode of the distribution increases with c. For the gamma distributions fitted for $I(0,0)$, c decreases relatively dramatically (compared to the other statistics) as a is increased (expanding the bandwidth).

The overall conclusion of this section is that the statistics of the beam converge extremely rapidly as the bandwidth of the turbulence is expanded in the direction of propagation. This confirms the accuracy of the Markov approximation for long propagation distances.

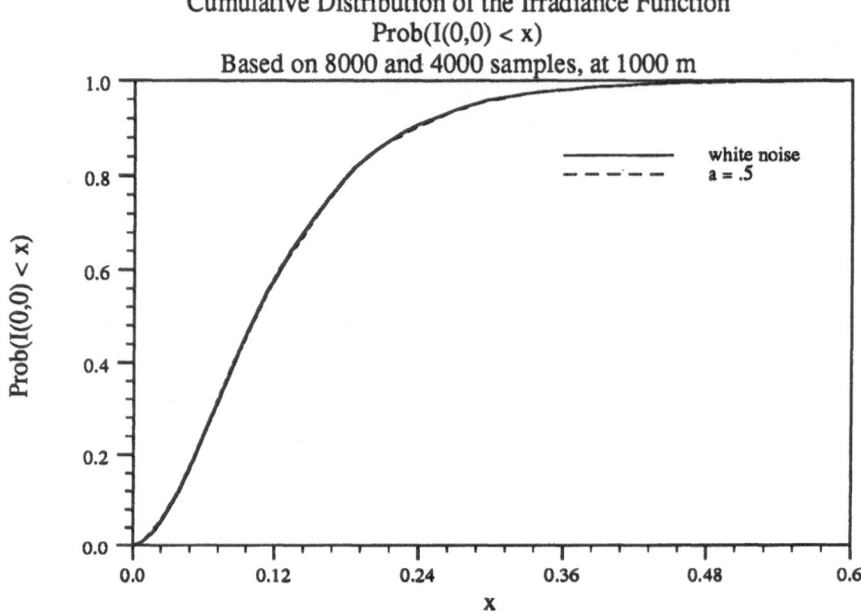

Figure 5.19: Cumulative Distribution Of $I(0,0)$ For White Noise And O-U Input $a = .5$

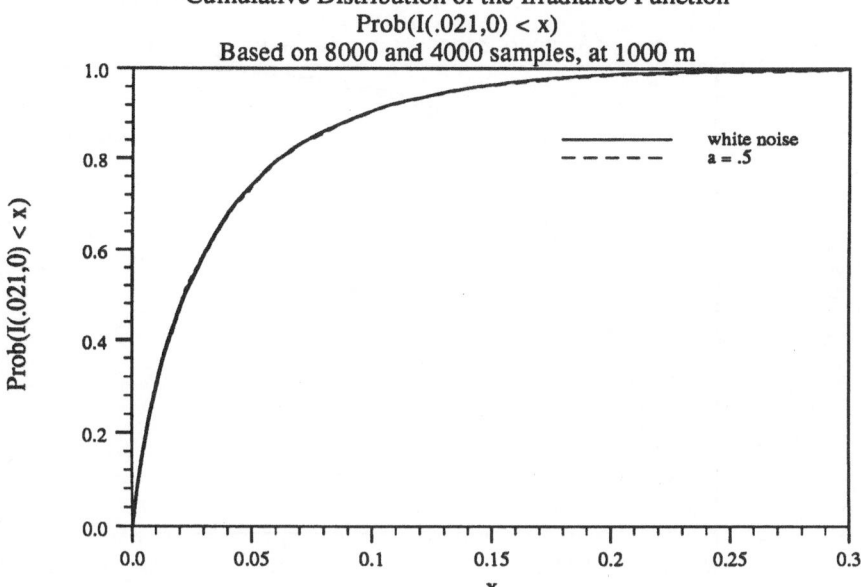

Figure 5.20: Cumulative Distribution Of $I(.021, 0)$ For White Noise And O-U Input $a = .5$

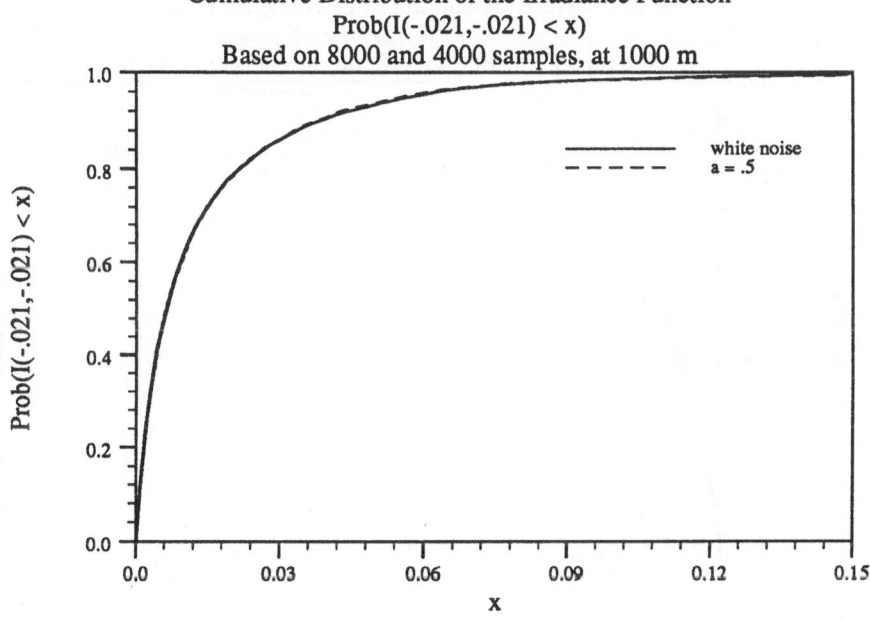

Figure 5.21: Cumulative Distribution Of $I(-.021, -.021)$ For White Noise And O-U Input $a = .5$

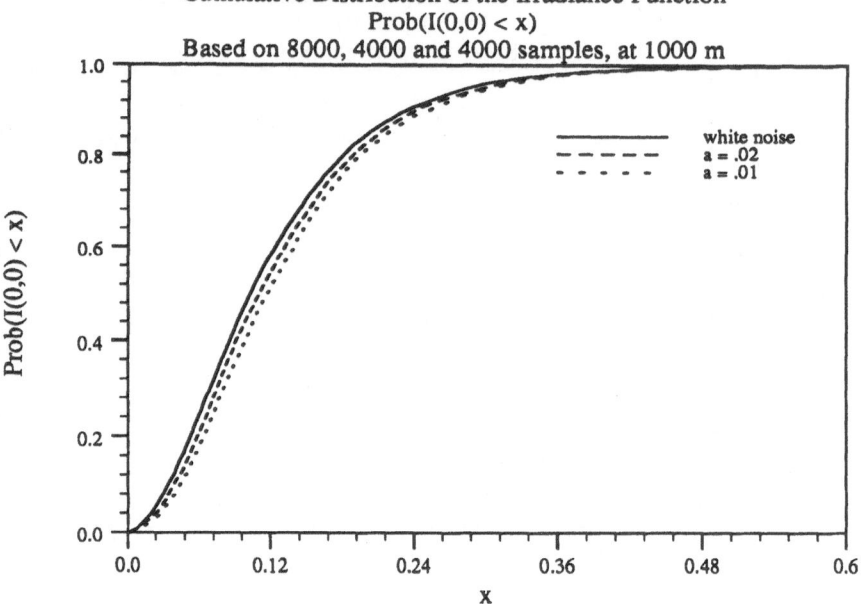

Figure 5.22: Cumulative Distribution Of $I(0,0)$ For White Noise And O-U Input $a = .02, .01$

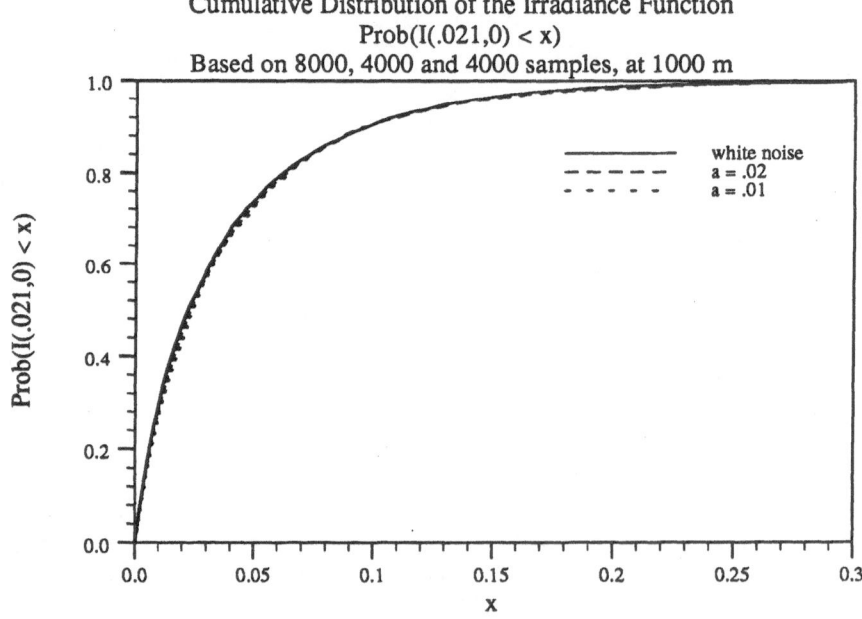

Figure 5.23: Cumulative Distribution Of $I(.021,0)$ For White Noise And O-U Input $a = .02, .01$

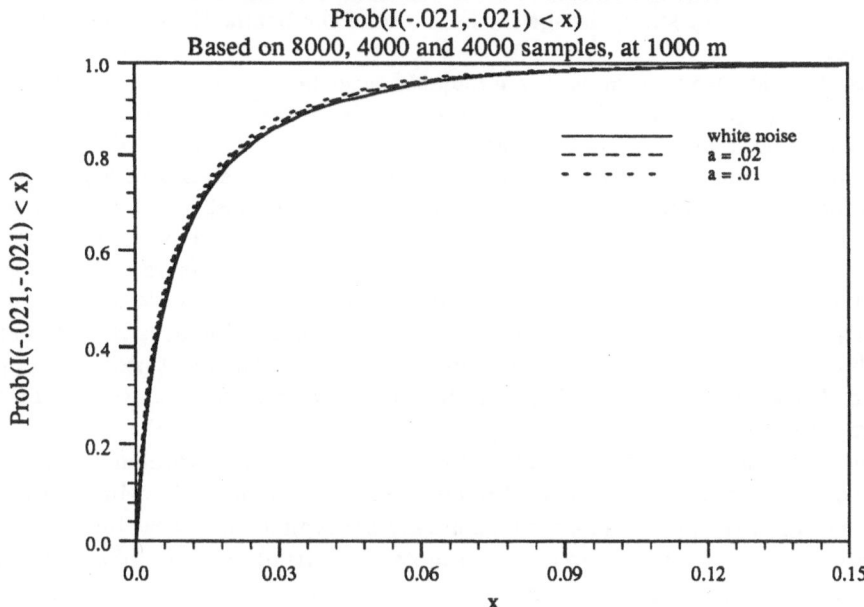

Figure 5.24: Cumulative Distribution Of $I(-.021, -.021)$ For White Noise And O-U Input $a = .02, .01$

5.9 Distortion Of The Beam

In this section I consider and illustrate some of the effects the atmosphere has on a laser beam. These are the results of the simulations using the statistical model for turbulence developed by Tatarskii and presented in Chapter 3. First I consider the irradiance and phase for an undistorted beam and then compare that to a typical simulated laser beam distorted by random turbulence.

In Figures 5.25 - 5.26 the irradiance function for a laser beam propagating in a vacuum is shown at 250 m, 500 m, 750 m and 1000 m. For the initial condition used here of

$$V_0(\rho) = e^{-|\rho|^2/10^{-4}}$$

and a wave number $k = 10^{-7}$, the peak value of the undistorted irradiance at z meters is $1/(1 + 4z^2 10^{-6})$. Hence the peak irradiance is 1. at 0 m, .8 at 250 m, .5 at 500 m, .4444 at 750 m and .2 at 1000 m.

In Figure 5.27 the same undistorted irradiance is displayed in a contour plot. The height of each contour here is .05. Note the spreading of the irradiance function as the beam propagates.

In Figures 5.28 - 5.29 the phase of the undistorted beam is shown as a mesh diagram. Note the pattern of wide variation with an island of calm in the center of the beam. The phase for the undistorted beam is also shown in Figure 5.30 in contour plot form.

The highly compressed bands of about 9 contours represent a change of about 2π, hence they represent a very small change in the phase of the beam. Each contour has a height of $2\pi/10$.

Next I consider an example of a simulated laser beam propagating in strong turbulence. Mesh graphs of the irradiance function are shown in Figures 5.31 - 5.32. Note the sharp peaking of the irradiance at 1000 m in contrast to the undistorted beam.

Contour graphs of the distorted irradiance function are shown in Figure 5.33 . The height for each contour is again .05. Comparing Figures 5.33 and 5.27 it is surprising to see that the peaks are higher for the distorted beam at 1000 m. A certain amount of beam splitting is evident at 1000 m, and the beam energy appears to be drifting away from the boresight, hence the beam is bending somewhat. As I increased the turbulence intensity, it was typical to see the bending of the beam increase dramatically, even though beam spreading was not greatly increased. Hence the most physically significant effect of turbulence on a laser beam as far as the irradiance is concerned may be the bending of the beam.

The irradiance for a beam distorted by a different typical turbulence field is given in Figure 5.34. For this second beam, the effect of the turbulence is primarily to increase the spreading of the beam. The peak values are not higher than those for the undistorted beam for this second simulation run. 3-D plots of the irradiance and phase of this second beam can be found in Appendix B.

The irradiance of a third beam distorted by turbulence is given in Figure 5.35. Note that in this case, the beam experiences hardly any spreading, and the effect of the turbulence is to focus and bend the beam, so that it moves away from the boresight. Note also the small value of the irradiance at the boresight.

The phase of the distorted beam is displayed in mesh graphs in Figures 5.36 - 5.37. The 'island of calm' seen in the center of the undistorted beam is now not so calm. This region, where the phase varies slowly, is also much smaller in size than it was for the undistorted beam.

Contour plots of the phase of the distorted beam are displayed in Figure 5.38. Unlike the phase of the undistorted beam, the phase is now highly unsymetric. The dark bands again indicate a change of 2π in the phase, which is really a small change.

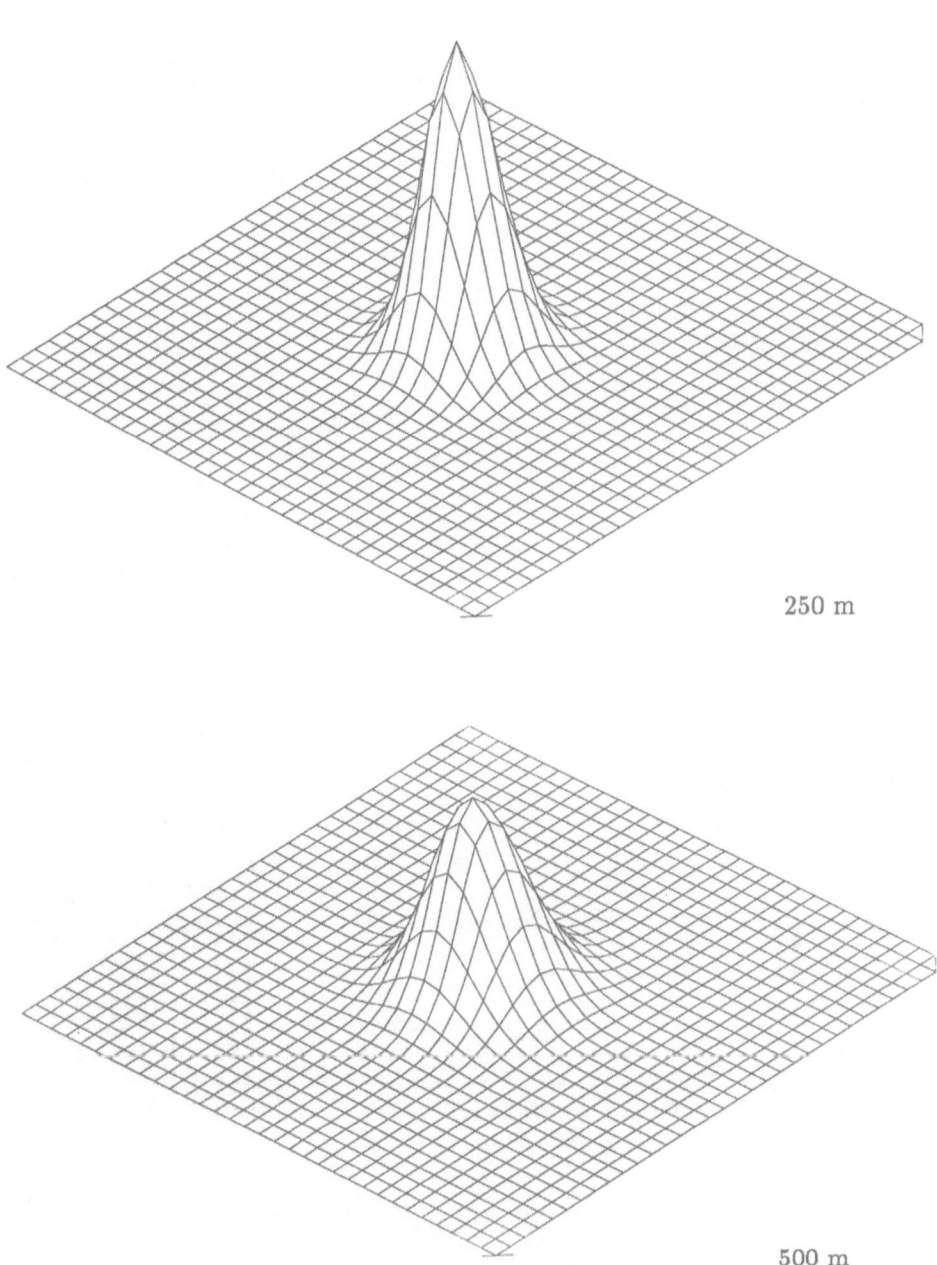

250 m

500 m

Figure 5.25: Undistorted Irradiance At 250 m And 500 m, 3-D Graph

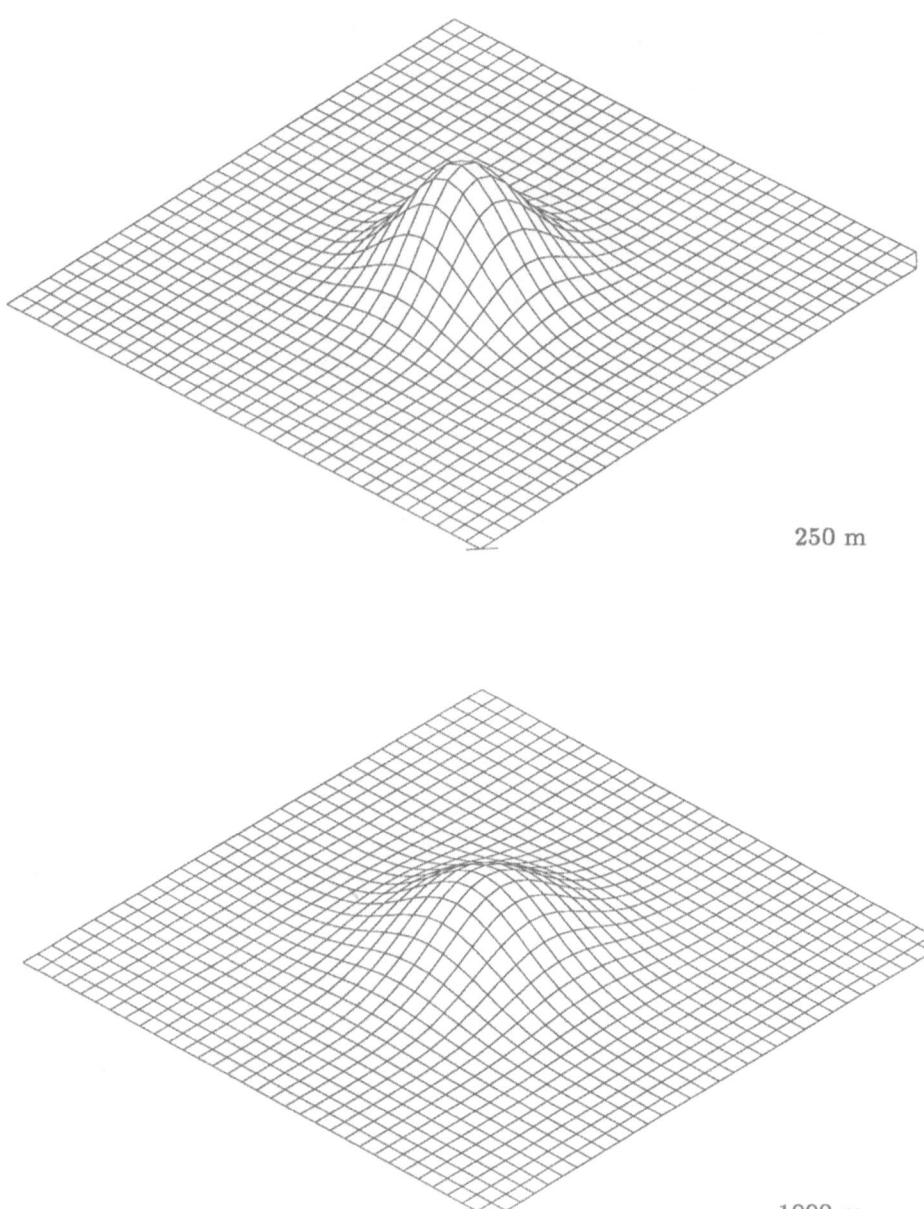

250 m

1000 m

Figure 5.26: Undistorted Irradiance At 750 m And 1000 m, 3-D Graph

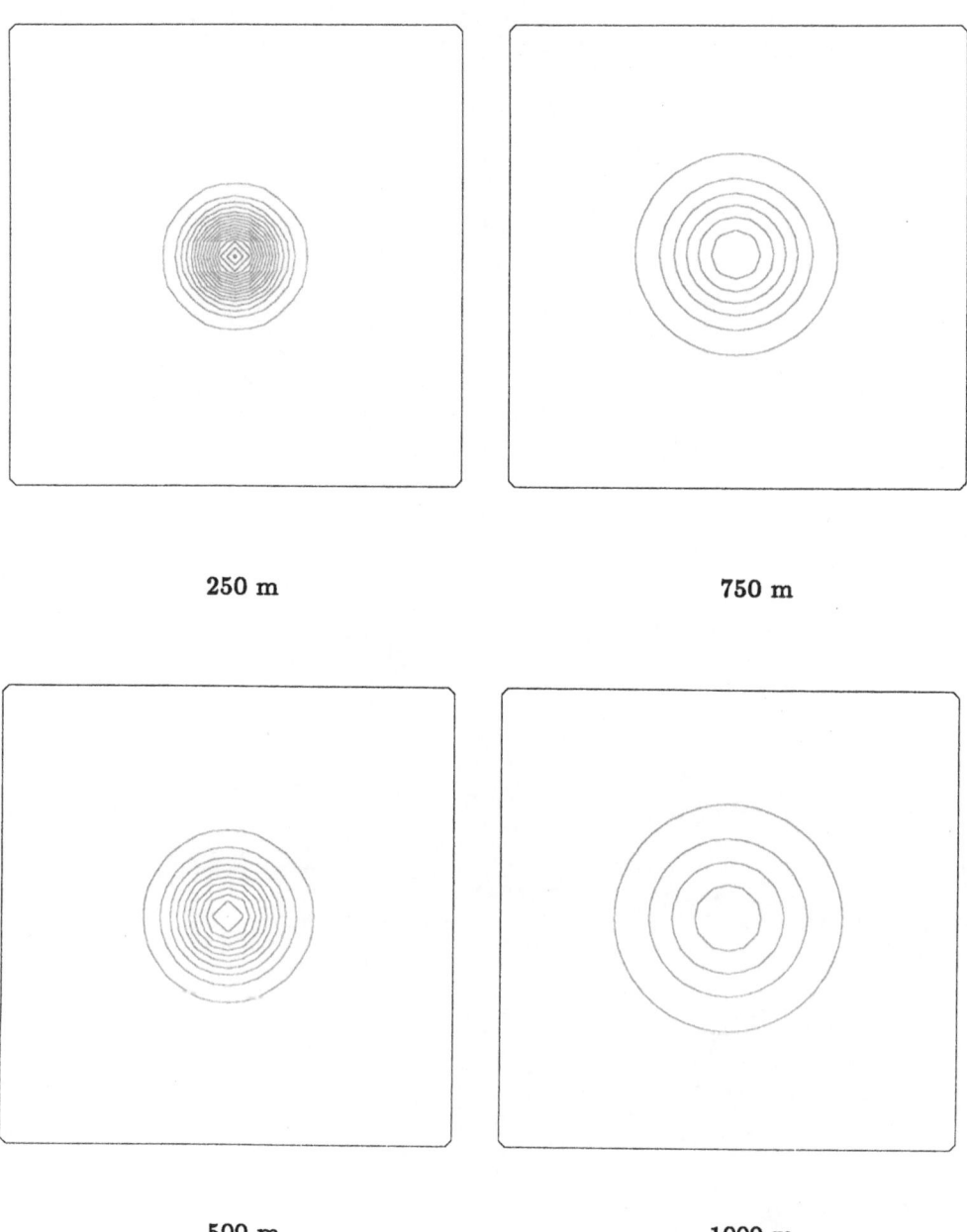

250 m   750 m

500 m   1000 m

Figure 5.27: Undistorted Irradiance, Contour Plot, step = .05

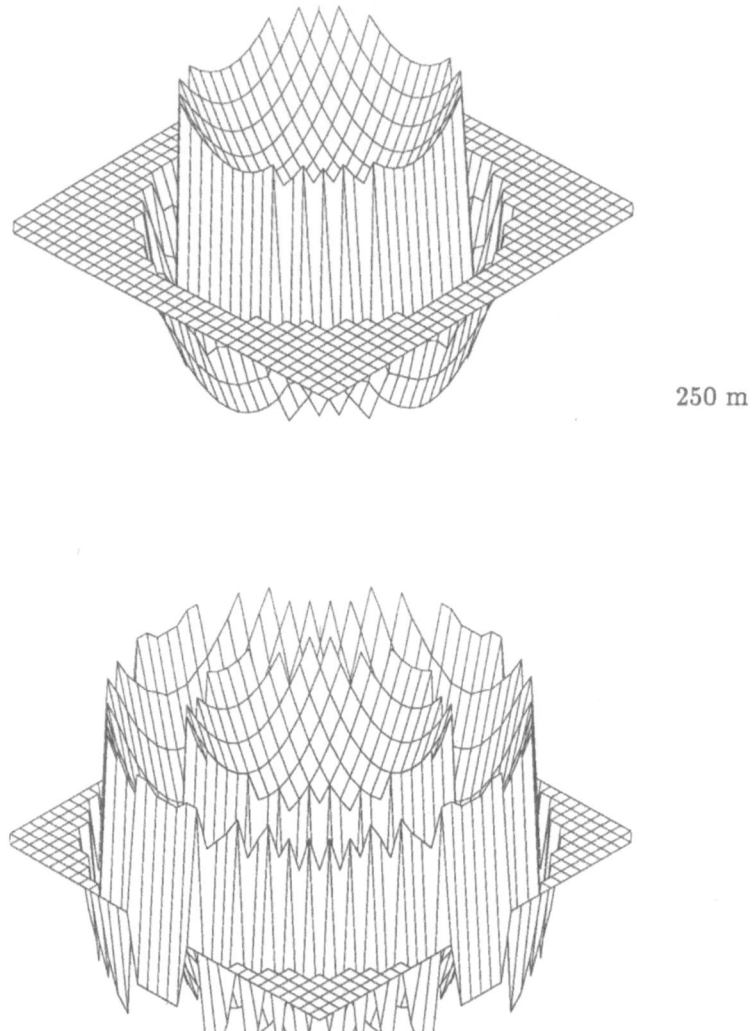

250 m

500 m

Figure 5.28: Undistorted Phase At 250 m And 500 m, 3-D Graph

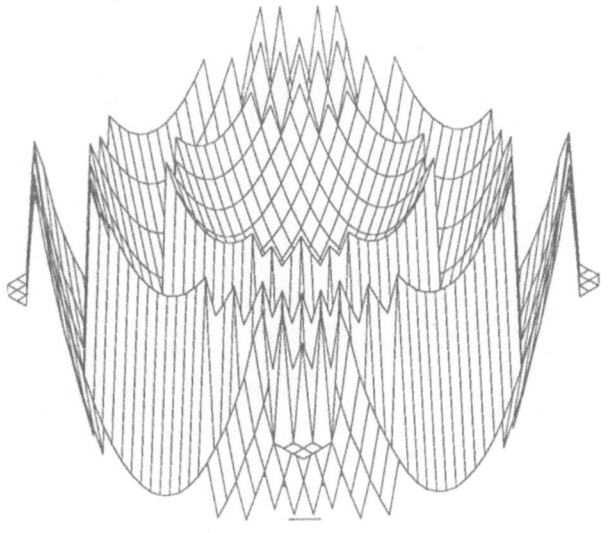

750 m

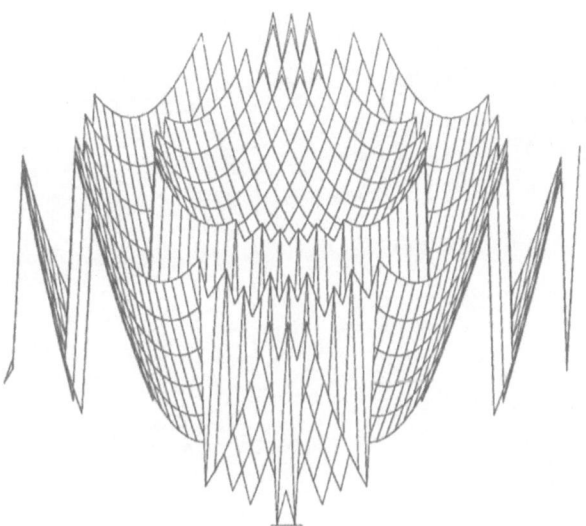

1000 m

Figure 5.29: Undistorted Phase At 750 m And 1000 m, 3-D Graph

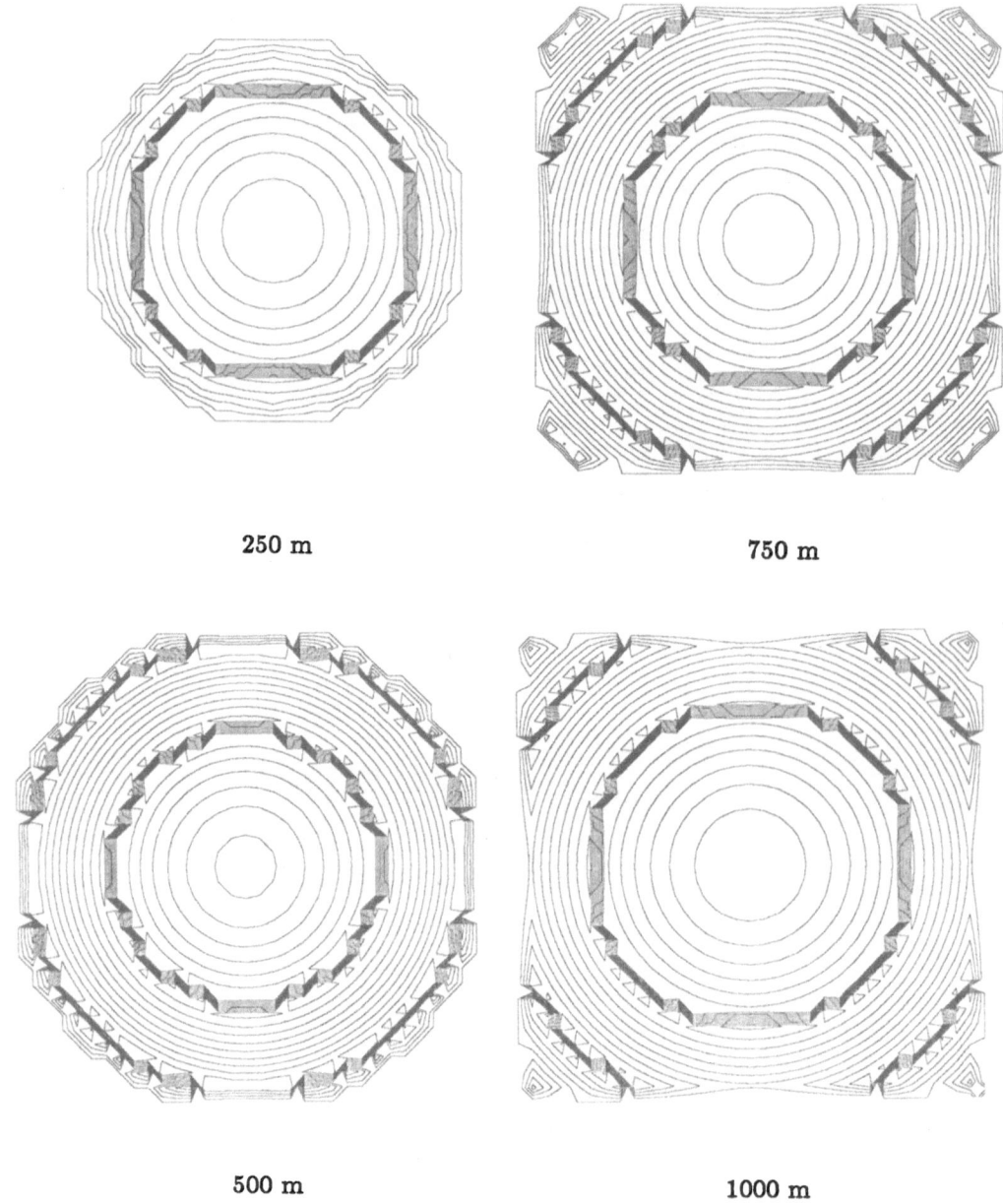

250 m   750 m

500 m   1000 m

Figure 5.30: Undistorted Phase, Contour Plot, step $= 2\pi/10$

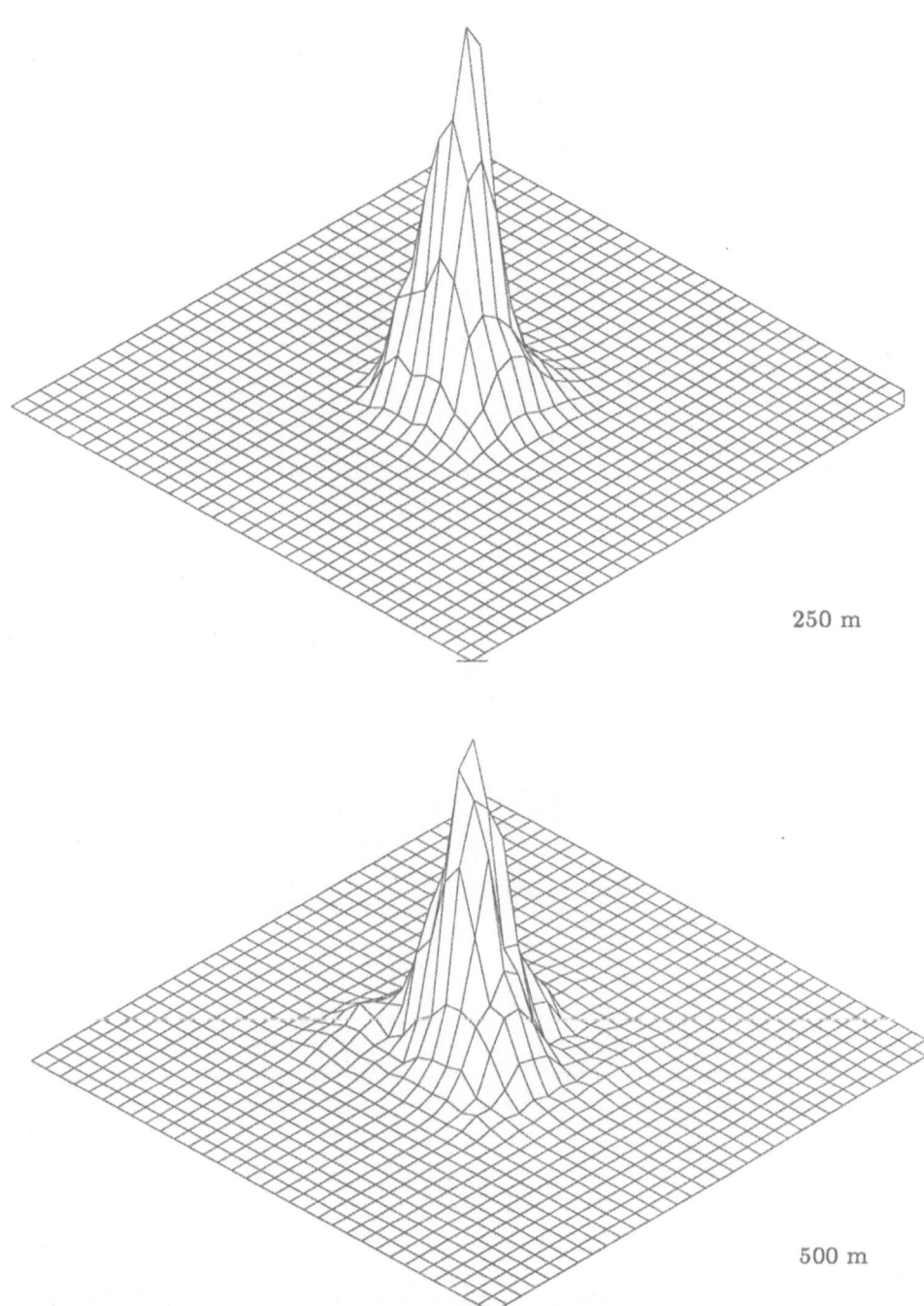

250 m

500 m

Figure 5.31: Distorted Irradiance At 250 m And 500 m, 3-D Graph, Run 1

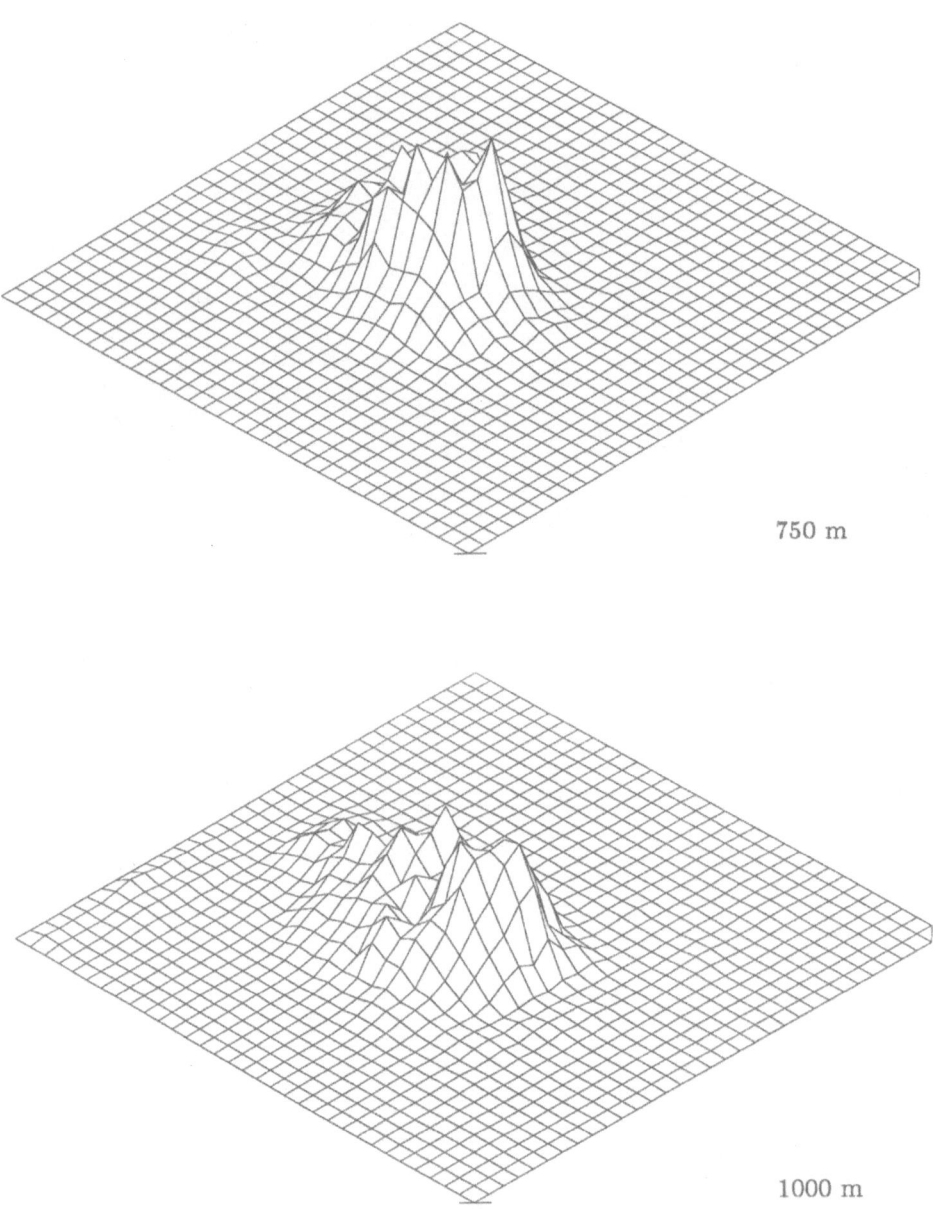

750 m

1000 m

Figure 5.32: Distorted Irradiance At 750 m And 1000 m, 3-D Graph, Run 1

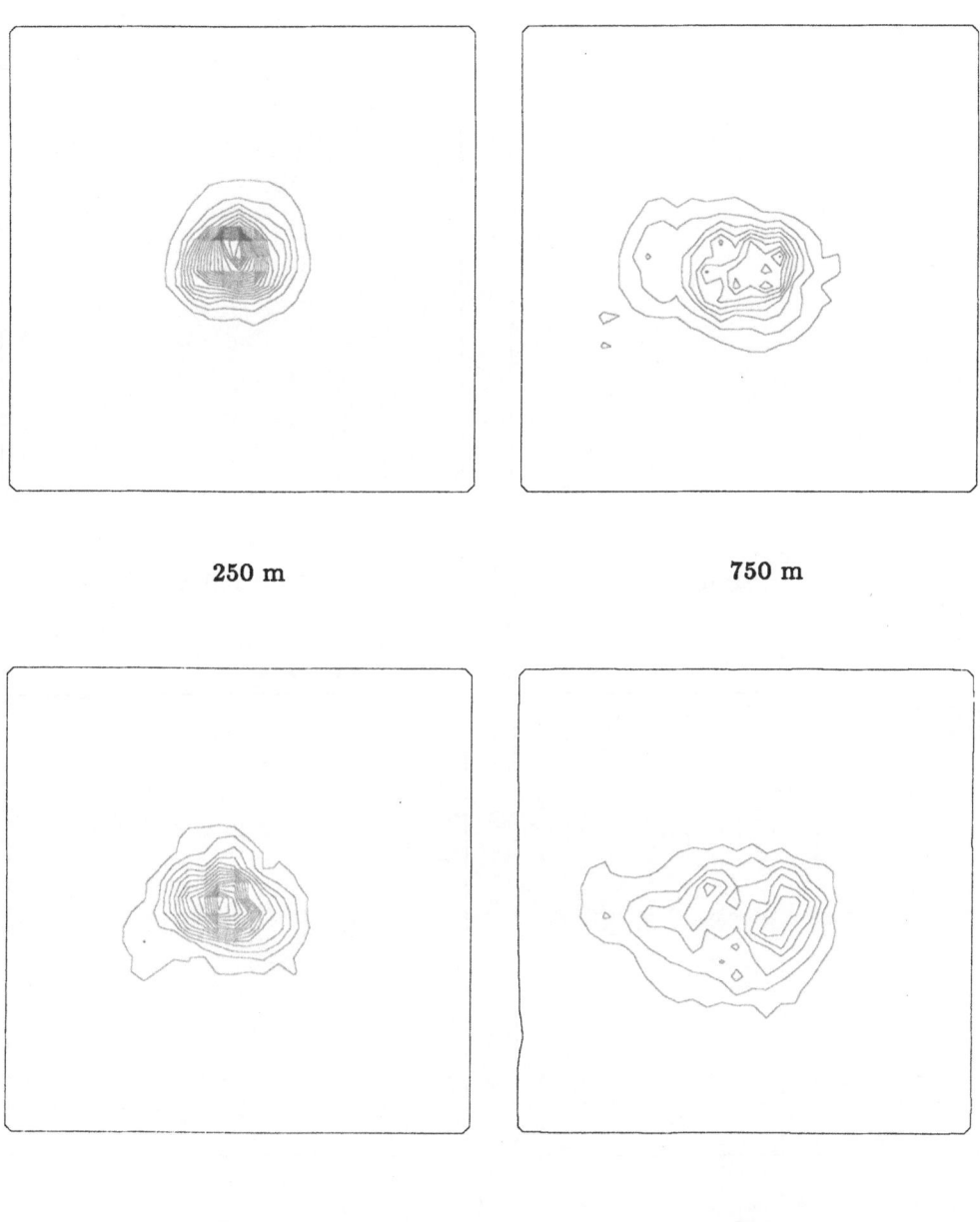

250 m

750 m

500 m

1000 m

Figure 5.33: Distorted Irradiance, Contour Plot, step = .05, Run 1

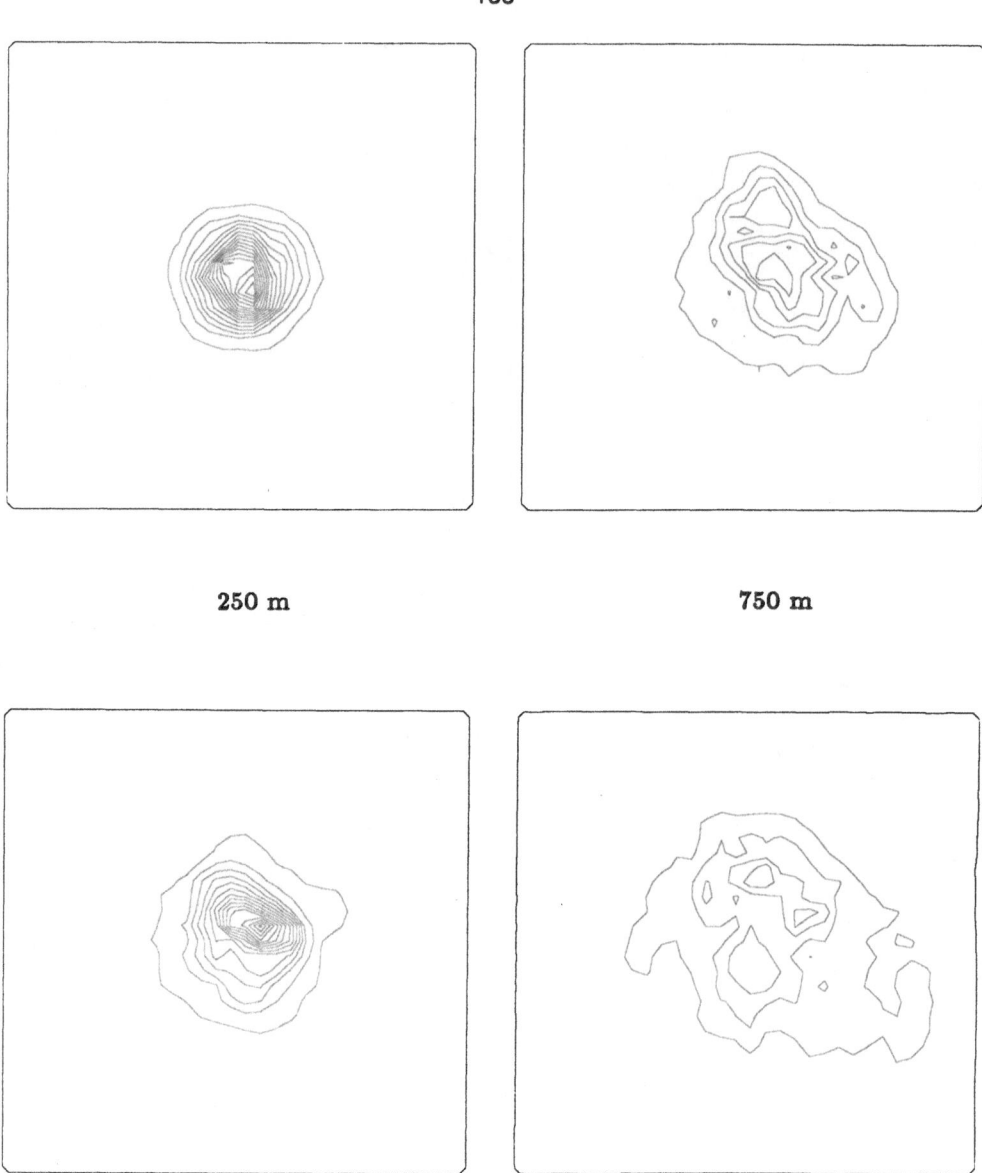

250 m   750 m

500 m   1000 m

Figure 5.34: Distorted Irradiance, Contour Plot, step = .05, Run 2

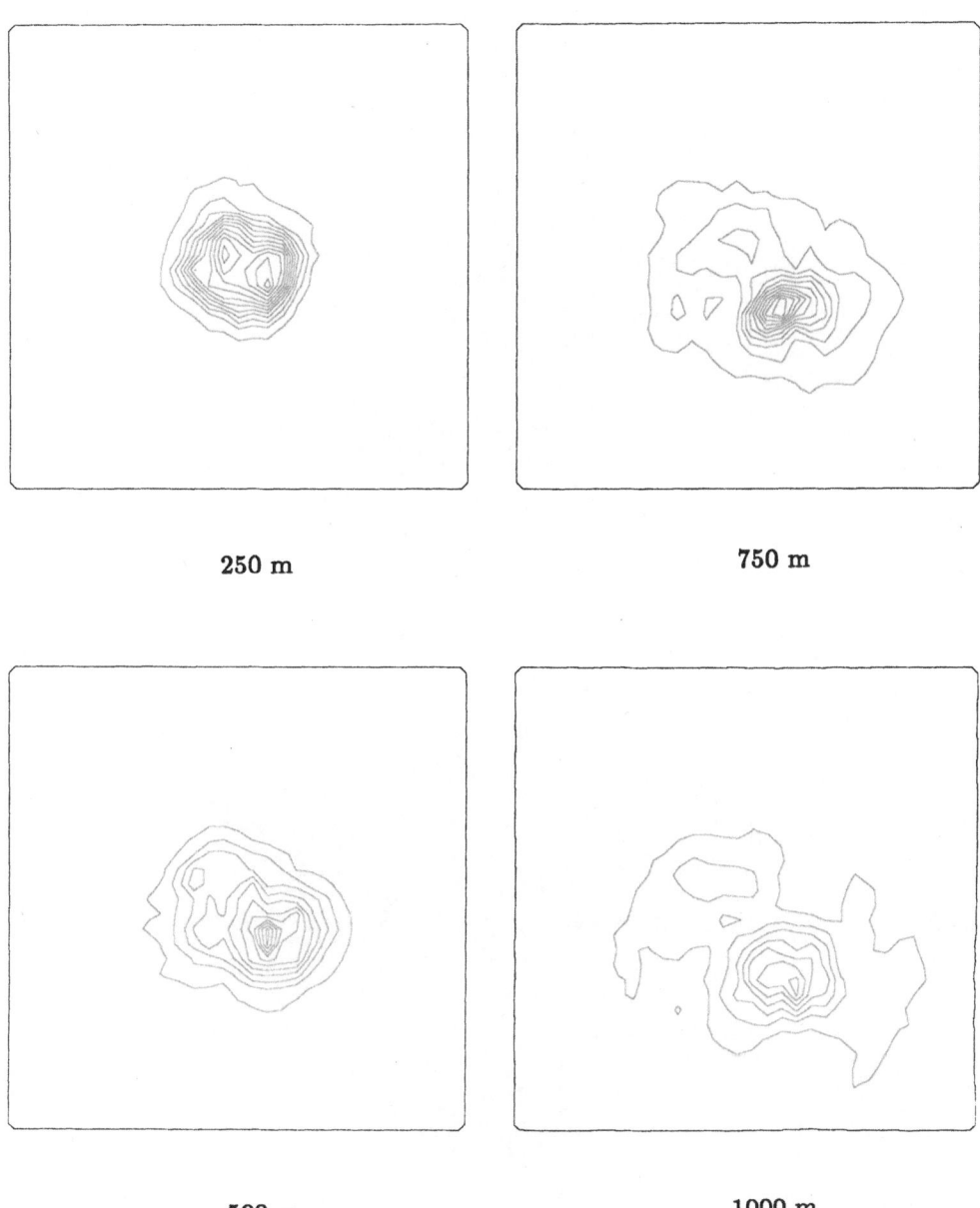

250 m

750 m

500 m

1000 m

Figure 5.35: Distorted Irradiance, Contour Plot, step = .05, Run 3

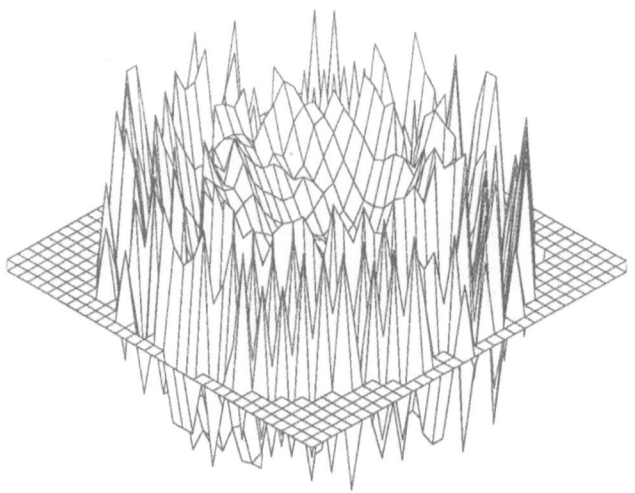

250 m

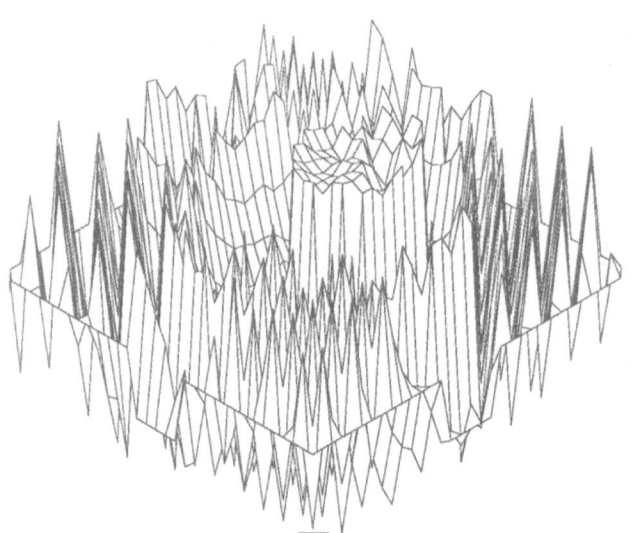

500 m

Figure 5.36: Distorted Phase At 250 m And 500 m, 3-D Graph, Run 1

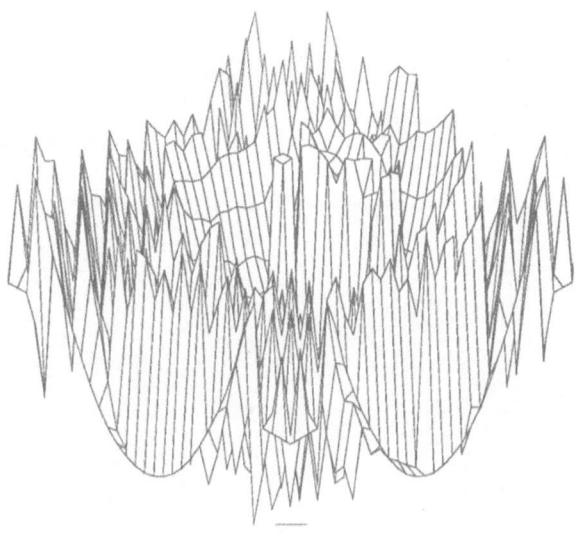

750 m

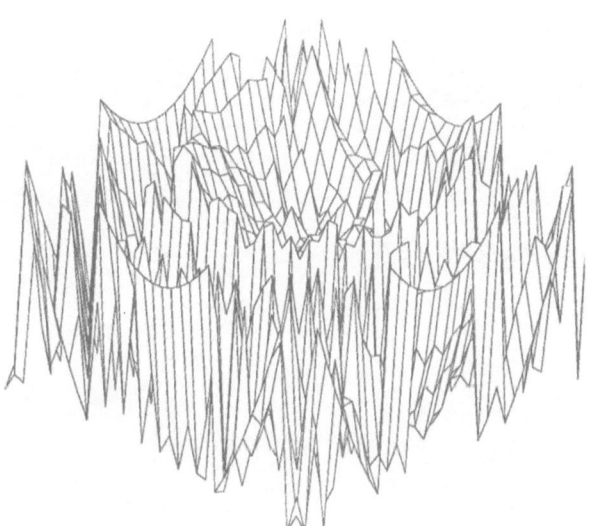

1000 m

Figure 5.37: Distorted Phase At 750 m And 1000 m, 3-D Graph, Run 1

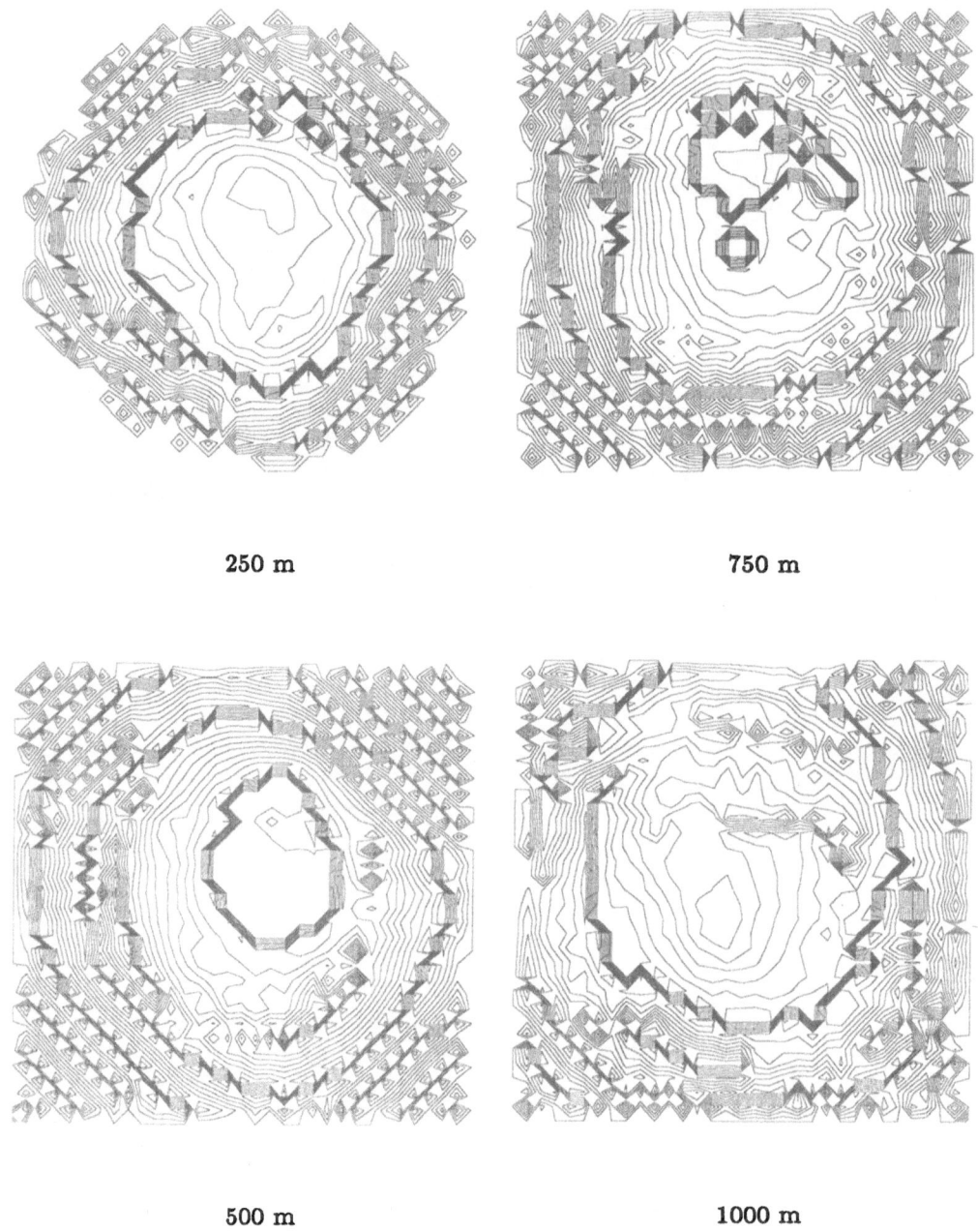

250 m

750 m

500 m

1000 m

Figure 5.38: Distorted Phase, Contour Plot, step $= 2\pi/10$, Run 1

Chapter 6

Feynman Path Integrals

The Feynman path integral is a powerful tool for investigating solutions of the Schroedinger equation, and in this chapter I use it to calculate some approximate solutions to the Forward Scattering Equation 2.19. It's relation to the product formulas developed in Chapter 4 is very clear and was part of Feynman's original approach to the problem. A fine mathematical discussion of Feynman integrals can be found in Albeverio [1]. A more recent paper by Kallianpur, Kannan and Karandikar [28] defines the Feynman integral in terms of abstract Wiener spaces and Hilbert spaces. Dashen [14] has also used the Feynman integral to compute moments for solutions to the forward scattering equation and explore other properties of the solutions.

6.1   Relation To Product Formulas

The purpose of this section is to make intuitive sense out of the Feynman integral and to relate it to the product forms of Chapter 4.

For a fixed t we can express the product form $V_t^l(\rho_0)$ as

$$
\begin{aligned}
V_t^l(\rho) &= \prod_{j=1}^{j=l}[\exp\{ik\int_{(j-1)t/l}^{jt/l} n_{1,\tau}\ d\tau\}S_{t/l}]V_0(\rho) \\
&= \int_{\mathbf{R}^{2l}}(\frac{1}{4\pi t/l})^l \prod_{j=1}^{l}[\exp\{\frac{ik}{2t/l}|\rho_j - \rho_{j-1}|^2\} \\
&\quad \times \exp\{\int_{(j-1)t/l}^{jt/l} n_{1,\tau}(\rho_{l-j+1})\ d\tau\}]V_0(\rho_l)\ d\rho \\
&= (\frac{1}{4\pi t/l})^l \int_{\mathbf{R}^{2l}} \exp\{\frac{ik}{2}\sum_{j=1}^{l}|(\rho^l(t_j) - \rho^l(t_{j-1}))/(t/l)|^2(t/l)\} \\
&\quad \times \exp\{ik\int_0^t n_{1,\tau}(\rho^l(t - \tau)\ d\tau\}V_0(\rho^l(t))\ d\rho
\end{aligned}
\tag{6.1}
$$

where

$$
\rho^l(s) = \rho_j, \ s \in [(j-1)t/l, jt/l)
$$

and

$$
t_j = jt/l
$$

If we let the function ρ^l converge to an absolutely continuous function , the integrand in Equation 6.1 converges to

$$\exp\{\frac{ik}{2}\int_0^t |\dot{\rho}(s)|^2 \, ds\} \exp\{ik \int_0^t n_{1,s}(\rho(t-s)) \, ds\} V_0(\rho(t)) \qquad (6.2)$$

It is quite reasonable to assume that ρ^l converges to a continuous function since the semigroup S_t is continuous at $t = 0$ hence it's Green's function behaves somewhat like a delta function. One of the assumptions of the Parabolic approximation was that for any path ρ the deflection angle was small, that is $|\rho(t)-\rho(s)|/(t-s)$ is also small and bounded . Therefore the assumption of an absolutely continuous ρ makes sense physically. Since the integral in Equation 6.1 was taken with respect to a translation invariant measure, the limiting integral should be also. Let $\dot{\rho}$ take on values in $H_t = L_2[(0,t); \mathbf{R}^2]$ and let μ be a translation invariant, finitely additive measure on that space. The limit of Equation 6.1 can be expressed as

$$V_t(\rho(0)) = \int_{H_t} \exp\{\frac{ik}{2}\|\dot{\rho}\|^2\} \exp\{ik \int_0^t n_{1,s}(\rho(t-s)) \, ds\} V_0(\rho(t)) \, d\mu(\dot{\rho}) \qquad (6.3)$$

which is the Feynman-Ito Equation for the solution to the forward scattering equation. The mathematical properties of this equation are explored in the next section.

6.2 A Path Integral For Laser Propagation: The Feynman-Ito Equation

In this section the existence of the integral in Equation 6.3 and it's correspondence to the solution of the forward scattering equation are discussed. First a few preliminary definitions are given.

Definition 6.2.1 *Let H be a separable Hilbert space. Denote by $\mathcal{F}[H]$ the set of all functions f on H such that*

$$f(h) = \int_H e^{i[h,g]} \, d\nu(g)$$

where ν is a complex valued measure on H with bounded variation. $\mathcal{F}[H]$ is called the Fresnel class on H. The Fresnel norm of such a function is $\|f\|_{\mathcal{F}} = |\nu|$.

The following Proposition is easily shown using the convolution properties of bounded measures, and it's proof is omitted.

Proposition 6.2.2 *The space $\mathcal{F}[H]$ is a Banach algebra.*

The Feynman integral

$$\int_H e^{\frac{i}{2}\|h\|^2} f(h) \, dh$$

exists when $f \in \mathcal{F}[H]$. If

$$f(h) = \int_H e^{i[h,g]} \, d\nu(g)$$

where ν is a bounded measure, then the Feynman integral of f is

$$\int_H e^{\frac{i}{2}\|h\|^2} f(h) \, dh = \int_H e^{\frac{-i}{2}\|g\|^2} \, d\nu(g)$$

It is also clear that

$$|\int_H e^{\frac{i}{2}\|h\|^2} f(h) \, dh| \leq \|f\|_{\mathcal{F}}$$

More detail on this can be found in Albeverio [1].

Now, suppose that $n_{1,t} = n_1$, i.e. it is constant in t. The following theorem, first established by Ito [26]

Theorem 6.2.3 *Suppose that V_0, $n_1 \in \mathcal{F}[\mathbf{R}^2]$, and n_1 is real valued. Then the solution to the forward scattering equation*

$$\dot{V}_t = \frac{i}{2k}\nabla^2 V_t + ik n_1 V_t$$

is given by the Feynman-Ito Equation 6.3.

Clearly this theorem can be trivially extended to the case where n_1 is a step function. Now the case of interest, where

$$n_{1,t}(\rho) = \int_{\mathbf{R}^2} w(\rho - \rho') N_t(\rho') \, d\rho'$$

and $N \in \mathcal{W} = L_2[(0,T); H]$ is considered.

Theorem 6.2.4 *Suppose that $V_0 \in \mathcal{F}[\mathbf{R}^2]$, $w \in L_2(\mathbf{R}^2)$ and $N \in \mathcal{W}$. Let $n_{1,s} = w \times N_s$. Then the solution to the forward scattering equation is given by Equation 6.3 and the integral in Equation 6.3 exists.*

Proof
As shown in Chapter 3, the solution to the forward scattering equation is a continuous function of N, hence N can be approximated by step functions. It is enough to show that the solution to Equation 6.3 exists and is a continuous function of N.

Denote $w_\rho(t, \rho') = w(\rho(t) - \rho')$. Then Equation 6.3 can be rewritten as

$$V_t(\rho(0)) = \int_{H_t} \exp\{\frac{ik}{2}\|\dot{\rho}\|^2\} \exp\{i[kw_\rho, N]\} V_0(\rho(t)) \, d\dot{\rho} \qquad (6.4)$$

Since $\mathcal{F}[H_t]$ is a Banach algebra, it is enough to show that $[w_\rho, N]$ and $V_0(\rho(t))$ are Fresnel class functions.

$$
\begin{aligned}
V_0(\rho(t)) &= \int_{\mathbf{R}^2} \exp\{i[\lambda, \rho(0)] + \int_0^t \dot{\rho}(s) \, ds\} \, d\nu(\lambda) \\
&= \int_{\mathbf{R}^2} \exp\{i[\lambda, \rho(0)]\} \exp\{i[\lambda 1_{[0,t]}, \dot{\rho}]\} \, d\nu(\lambda)
\end{aligned}
$$

and $\nu \times m$ where m is a unit point mass measure on $L_2[0,t]$ at $h = 1$, is a bounded measure on H_t. Hence $V_0(\rho(t))$ is Fresnel class.

Denote

$$f_t(h) = \int_{\mathbf{R}^2} \exp\{i[h, \lambda]\} \hat{f}_t(\lambda) \, d\lambda$$

Then we can write

$$
\begin{aligned}
[kw_\rho, N] &= k \int_0^t \int_{\mathbf{R}^2} \exp\{i[\lambda, \rho(t-s)]\} \hat{w}(\lambda) \hat{N}_s(\lambda) \, d\lambda \, ds \\
&= k \int_0^t \int_{\mathbf{R}^2} \exp\{i[\lambda, \rho(0)]\} \exp\{i[\lambda 1_{[0,t-s]}, \dot{\rho}]\} \hat{w}(\lambda) \hat{N}_s(\lambda) \, d\lambda \, ds
\end{aligned}
$$

And the measure on H_t defined by $|\hat{N}_s(\lambda)\hat{w}(\lambda)| m_s$ where m_s is a point mass at $1_{[0,t-s]}$. Since w and N are square integrable, this measure is also finite and its variation is less than $k\sqrt{t}\|w\|\|N\|$. Because of the Banach algebra property

$$\| \exp\{i[w_\rho, N]\} V_0(\rho(t)) \|_{\mathcal{F}} \leq \exp\{k\sqrt{t}\|w\|\|N\|\} \|V_0\|_{\mathcal{F}}$$

Hence the integral exists. To show continuity, consider

$$
\begin{aligned}
\| \exp\{i[w_\rho, N_1]\} &- \exp\{ik[w_\rho, N_2]\} \|_{\mathcal{F}} \leq \\
&\leq \| \exp\{ik[w_\rho, N_2]\} \|_{\mathcal{F}} \| \exp\{ik[w_\rho, N_1 - N_2]\} - 1 \|_{\mathcal{F}} \\
&\leq \exp\{k\sqrt{t}\|w\|\|N_2\|\} (\exp\{k\sqrt{t}\|w\|\|N_1 - N_2\|\} - 1)
\end{aligned}
$$

Hence the Feynman-Ito integral is a continuous function of N. Hence the solution to the forward scattering equation is given by Equation 6.3. $\square$

In the next six sections, the Feynman integral is used to derive several useful approximations and gives some insight into what is happening as the laser beam passes through the atmosphere.

6.3   Discussion Of The Work Of K. Furutsu

K. Furutsu [19] was able to calculate many statistics for the beam irradiance $I(\rho) = |V(\rho)|^2$ by assuming a structure function for $n_{1,t}$ of

$$D(\rho) = |\rho|^2$$

It was later pointed out by Tatarskii and others that this was equivalent to assuming that $n_{1,t}$ was linear in ρ and could be represented as

$$n_{1,t}(\rho) = [N_t, \rho]$$

where N is an $L_2[(0,T); \mathbf{R}^2]$ valued independent white noise. Denote $W_t = \int_0^t N_s \, ds$ The Feynman integral for the solution corresponding to this turbulence field is

$$
\begin{aligned}
V_t(\rho(0)) &= \\
&= \int_{H_t} \exp\{\frac{ik}{2}\|\dot{\rho}\|^2\} \exp\{ik \int_0^t [\rho(s), N_{t-s}] \, ds\} V_0(\rho(t)) \, d\dot{\rho} \\
&= \int_{H_t} \exp\{\frac{ik}{2}\|\dot{\rho}\|^2\} \exp\{ik([\rho(0), W_t] + \int_0^t [\dot{\rho}(s), W_{t-s}] \, ds\} V_0(\rho(0) + \int_0^t \dot{\rho}(s) \, ds) \, d\dot{\rho}
\end{aligned}
$$

$$= \int_{H_t} \exp\{\frac{ik}{2}\|\dot{\rho} + W_{t-\cdot}\|^2\} \exp\{ik[\rho(0), W_t]\} \exp\{\frac{-ik}{2}\|W\|^2\} V_0(\rho(0) + \int_0^t \dot{\rho}(s) \, ds) \, d\dot{\rho}$$

$$= \int_{H_t} \exp\{\frac{ik}{2}\|\dot{\rho}\|^2\} \exp\{\frac{ik}{2}\|W\|^2 + ik[\rho(0), W_t]\} V_0(\rho(t) - \int_0^t W_{t-s} \, ds) \, d\dot{\rho}$$

$$= \exp\{\frac{ik}{2}\|W\|^2 + ik[\rho(0), W_t]\} [S_t V_0](\rho(0) - \int_0^t W_s \, ds) \, d\dot{\rho}$$

since the measure on H_t is translation invariant.

Hence this solution corresponds to a random phase shift as well as bending of the beam, or a spatial shift of the beam intensity. The wavefront is tilted as well. Obviously this is not an adequate description of the effect of turbulence on a laser beam, but it gives some insight into what is happening. A linear trend in the index of refraction results in a bending of the beam and tilting of the wavefront.

In their paper, Dawson and Papanicolaou [16] consider a limiting spectral density for $n_{1,t}$ of $\Phi(\lambda) = \sigma_\epsilon p_\epsilon(\lambda)$, where

$$\int_{\mathbf{R}^2} p_\epsilon(\lambda) \, d\lambda = 1$$

$$\sigma_\epsilon^2 = \sigma^2/\epsilon$$

$$\int_{\mathbf{R}^2} |\lambda|^2 p_\epsilon(\lambda) \, d\lambda = 1/\sigma_\epsilon^2$$

$$\sigma_\epsilon^2 \int_{\mathbf{R}^2} |\lambda|^4 p_\epsilon(\lambda) \, d\lambda = O(\epsilon)$$

and consider the behavior of V_t as $\epsilon \downarrow 0$. The structure function for these spectral densities is

$$D(\rho) = |\rho|^2/2 + O(\epsilon\rho^4)$$

Hence the limiting structure function is $D(\rho) = |\rho|^2/2$ which is identical to Furutsu's assumption.

6.4   First Order Approximate Solutions

In this section a modification to Furutsu's result is considered. Instead of assuming the turbulence field is linear, it is assumed to be so only locally. I make the approximation

$$n_{1,s}(\rho') \approx n_{1,s}(\rho(0)) + [\nabla n_{1,s}(\rho(0)), \rho' - \rho(0)]$$

Once again, denote $W_t = \int_0^t n_{1,s} \, ds$. Plugging this approximation into the Feynman integral yields

$$V_t(\rho(0)) =$$

$$= \int_{H_t} \exp\{\frac{ik}{2}\|\dot{\rho}\|^2\} \exp\{ik \int_0^t n_{1,s}(\rho(t-s)) \, ds\} V_0(\rho(t)) \, d\dot{\rho}$$

$$\approx \int_{H_t} \exp\{\frac{ik}{2}\|\dot{\rho}\|^2\} \exp\{ik \int_0^t n_{1,s}(\rho(0)) + [\nabla n_{1,s}(\rho(0)), \rho(t-s) - \rho(0)] \, ds\} V_0(\rho(t)) \, d\dot{\rho}$$

$$= \int_{H_t} \exp\{\frac{ik}{2}\|\dot{\rho}\|^2\} \exp\{ik W_t(\rho(0)) + ik \int_0^t [\nabla W_{t-s}(\rho(0)), \dot{\rho}(s)] \, ds\} V_0(\rho(t)) \, d\dot{\rho}$$

$$= \int_{H_t} \exp\{\frac{ik}{2}\|\dot{\rho}\|^2\} \exp\{ikW_t(\rho(0)) - \frac{ik}{2}\|\nabla W(\rho(0))\|^2\}V_0(\rho(t) - \int_0^t \nabla W_s(\rho(0)) \ ds) \ d\dot{\rho}$$

$$= \exp\{ikW_t(\rho(0)) - \frac{ik}{2}\|\nabla W(\rho(0))\|^2\}[S_tV_0](\rho(0) - \int_0^t \nabla W_s(\rho(0)) \ ds)$$

taking advantage of the translation invariant property as in the previous section.

This approximation is an improvement on Furutsu's result because it takes into account not only the beam bending effect, but also the distortion of the shape and phase of the beam. The distribution of the irradiance function will be the same for this approximation as in Furutsu's result, so this is not a good question to address using this method. Also , as in Furutsu's result, this method yields a constant times a random phase in the plane wave case $V_0 \equiv 1$, and hence is only applicable in the beam wave case.

6.5   Locally Linear Approximate Solutions

In the previous section a locally linear approximation of n_1 was made using the Taylor series expansion. Since taking gradients may be undesirable and is difficult to do for a discretized computer simulation , it is helpful to note that any local linearization of n_1 will suffice. For example we can use the linearization corresponding to

$$\min \int_{\rho(0)+A} [n_{1,t}(\rho') - n_{m,t}(\rho(0)) - [n_{L,t}(\rho(0)), \rho' - \rho(0)]]^2 \ d\rho'$$

where A is a neighborhood of the origin. The linearization is then

$$n_{1,t}(\rho') = n_{m,t}(\rho(0)) + [n_{L,t}(\rho(0)), \rho' - \rho(0)]$$

where

$$n_{m,t}(\rho(0)) = \frac{1}{|A|} \int_{A+\rho(0)} n_{1,t}(\rho') \ d\rho'$$

$$n_{L,t}(\rho(0)) = [\int_{\rho(0)+A} (\rho' - \rho(0))(\rho' - \rho(0))^* \ d\rho']^{-1}$$
$$\times \int_{A+\rho(0)} (n_{1,t}(\rho') - n_{m,t}(\rho(0)))(\rho' - \rho(0)) \ d\rho'$$

Then if

$$W_{L,t}(\rho(0)) = \int_0^t n_{L,s}(\rho(0)) \ ds$$

$$W_{m,t}(\rho(0)) = \int_0^t n_{m,s}(\rho(0)) \ ds$$

we have

$$V_t(\rho(0)) = \exp\{ikW_{m,t}(\rho(0)) - \frac{ik}{2}\|W_L(\rho(0))\|^2\}[S_tV_0](\rho(0) - \int_0^t W_{L,s}(\rho(0)) \ ds) \quad (6.5)$$

and this is the most general form of the locally linearized solution. Note that differentiability is not required to use this local linearization method.

6.6 Second Order Approximate Solutions

The first order approximation methods given in the last two sections obviously have their limitations. Their accuracy degrades rapidly over longer propagation distances t. They are also not applicable to the plane wave case. For these reasons a second order approximation is now considered.

Let

$$
\begin{aligned}
n_{1,t}(\rho') &\approx n_{1,t}(\rho(0)) + [n_{L,t}(\rho(0)), \rho' - \rho(0)] \\
&\quad + \frac{1}{2}[n_{2,t}(\rho(0))(\rho' - \rho(0)), \rho' - \rho(0)]
\end{aligned}
\tag{6.6}
$$

Then

$$
\begin{aligned}
\int_0^t n_{1,s}(\rho(t-s))\ ds &= \int_0^t n_{m,s}(\rho(0)) + [n_{L,s}(\rho(0)), \rho(t-s) - \rho(0)] \\
&\quad + \frac{1}{2}[n_{2,s}(\rho(t-s) - rho(0)), \rho(t-s) - \rho(0)]\ ds \\
&= W_{m,t}(\rho(0)) + [W_{L,t-\cdot}(\rho(0)), \dot\rho] + \frac{1}{2}[R\dot\rho, \dot\rho]
\end{aligned}
$$

where $R : H_t \to H_t$ is defined as

$$
[Rf]_s = \int_0^t \left(\int_0^{t-s(\vee\sigma)} n_{2,r}\ dr\right) f_\sigma\ d\sigma
$$

It is assumed that the values of n_1 and n_2 are very small, hence $I + R$ is assumed to have bounded inverse and finite determinant. If n_2 is uniformly bounded then R is a nuclear operator and $\det(I+R)$ exists. Hence we can express V_t as

$$
V_t(\rho(0)) =
$$
$$
\begin{aligned}
&= \int_{H_t} \exp\{\frac{ik}{2}\|\dot\rho\|^2\} \exp\{ik(W_{m,t}(\rho(0)) + [W_{L,t-\cdot}(\rho(0)), \dot\rho] + \frac{1}{2}[R\dot\rho, \dot\rho])\} V_0(\rho(t))\ d\dot\rho \\
&= \int_{H_t} \exp\{\frac{ik}{2}[(I+R)\dot\rho, \dot\rho]\} \exp\{ik(W_{m,t}(\rho(0)) + [(I+R)(I+R)^{-1}W_{L,t-\cdot}(\rho(0)), \dot\rho]\} \\
&\quad \times V_0(\rho(t))\ d\dot\rho \\
&= \int_{H_t} \exp\{\frac{ik}{2}[(I+R)\dot\rho, \dot\rho]\} \exp\{ikW_{m,t}(\rho(0)) - \frac{ik}{2}[(I+R)^{-1}W_{L,t-\cdot}(\rho(0)), W_{L,t-\cdot}(\rho(0))]\} \\
&\quad \times V_0(\rho(t) - \int_0^t [(I+R)^{-1}W_{L,t-\cdot}(\rho(0))]_s\ ds)\ d\dot\rho
\end{aligned}
$$

For the beam wave case this is as far as we can go. In the plane wave case, in which all shifting of the beam is ignored, it is possible to obtain an expression for V_t and $I_t = |V_t|^2$. Kallianpur, Kannan and Karandikar [28] derived a Cameron-Martin formula for the Feynman-Ito integral which is given in the theorem below.

Theorem 6.6.1 *Let A be a self-adjoint nuclear operator on a separable Hilbert space H, such that $(I+A)^{-1}$ exists. Let f be a Fresnel class function on H with*

$$
f(x) = \int_H e^{i[x,h]}\ d\nu(h)
$$

Denote by $ind(I + A)$ *the number of negative eigenvalues of* $I + A$. *Then*

$$\int_H \exp\{\frac{i}{2}[(I+A)h,h]\}f(h) \; dh =$$

$$= \; |\det(I+A)|^{-\frac{1}{2}} \exp\{\frac{-i\pi}{2}(ind(I+A))\} \int_H \exp\{\frac{i}{2}[(I+A)^{-1}h,h]\} \; d\nu(h) \quad (6.7)$$

Note that our assumptions for R imply that $ind(I+R) = 0$. In the plane wave case our measure ν is simply a unit point mass at zero. Hence V_t can be expressed

$$V_t(\rho) = \exp\{ik(W_t(\rho) - \frac{1}{2}[(I+R)^{-1}W_{L,t-\cdot}(\rho),W_{L,t-\cdot}(\rho)]\}|\det(I+R)|^{-\frac{1}{2}} \quad (6.8)$$

And I_t can be expressed as

$$I_t(\rho) = |\det(I+R)|^{-1} \quad (6.9)$$

It remains to solve for $\det(I+R)$. Let

$$[I^- f](s) = \int_0^{t-s} f_\sigma \; d\sigma$$

Note that I^- is self adjoint and Hilbert-Schmidt on the space $L_2[(0,t);\mathbf{R}^n]$. Then R can be written as

$$R = I^- n_2 I^-$$

and for any uniformly bounded n_2, R is nuclear.

The calculation of the determinant is facilitated by performing a similarity transform on R. Let $\tilde{R} = T^*RT$ where T is the flip operator defined as $(Tf)_s = f_{t-s}$. Then

$$\tilde{R} = I^+ n_2 I^{+*}$$

where $(I^+ f)_s = \int_0^s f_\tau \; d\tau$ and

$$\det(I+R) = \det(I+\tilde{R})$$

From Balakrishnan [8] we know that

$$\log \det(I+\tilde{R}) \;=\; Tr. \; \log(I+\tilde{R})$$

$$=\; \sum_{k=1}^{\infty} \frac{(-1)^{k+1}}{k} Tr. \; \tilde{R}^k$$

For each k, $\tilde{R}^k$ is trace class of course, hence its trace can be calculated from its kernel $R_k(s,r)$. Now

$$(\tilde{R}f)_s = \int_0^t \int_0^{s \wedge r} n_2(\sigma) \; d\sigma f_r \; dr$$

hence

$$R_k(s,r) = (\int_0^t)^{k-1} \prod_{j=1}^{k} (\int_0^{s_{j-1} \wedge s_j} n_2(\sigma) \; d\sigma) \; ds_1 \cdots ds_{k-1}$$

taking $s_0 = s$ and $s_n = r$. Hence

$$
\begin{aligned}
Tr. \, \tilde{R}^k &= (\int_0^t)^k tr. \, \prod_{j=1}^{k-1} (\int_0^{s_{j-1} \wedge s_j} n_2(\sigma) \, d\sigma) \\
&\quad \times \int_0^{s_k \wedge s_0} n_2(\sigma) \, d\sigma \, ds_0 \cdots ds_k \\
&= (\int_0^t)^k tr. \, [\prod_{j=1}^{k-1} (t - (s_j \vee s_{j+1})) n_2(s_j)] n_2(s_k)(t - (s_k \vee s_1)) \, ds_1 \cdots ds_k
\end{aligned}
$$

Hence the expression for the irradiance function is

$$
\log I_t = - \sum_{k=1}^{\infty} (\int_0^t)^k [\prod_{j=1}^{k-1} (t - (s_j \vee s_{j+1}))](t - (s_k \vee s_1)) tr. \, n_2(s_1) \cdots n_2(s_k) \, ds_1 \cdots ds_k \quad (6.10)
$$

If only the first term in the sum is kept and n_2 is assumed to be Gaussian then I_t is log-normal. Once again, this approximation will only be valid over very small distances, however it provides insight into how small scale variations in the turbulence field effect the irradiance function.

6.7 Approximate First Moment

One thing the Feynman integral allows us to do is to take expectations in a fairly straightforward way. The only random term is

$$
\exp\{ik \int_0^t n_{1,s}(\rho(t - s)) \, ds\}
$$

In this section I consider an approximation expression for the mean field without making use of the Markov approximation. Hence, n_1 is assumed to be an isotropic homogeneous random field on $\mathbf{R}^3$ with covariance K and spectral density Φ.

Assume that it is possible to interchange expectation and the Feynman integral. Then

$$
\begin{aligned}
E[V_t(\rho(0))] &= E \int_{H_t} \exp\{\frac{ik}{2}\|\dot{\rho}\|^2\} \exp\{ik \int_0^t n_{1,s}(\rho(t - s) \, ds\} V_0(\rho(t)) \, d\dot{\rho} \\
&= \int_{H_t} \exp\{\frac{ik}{2}\|\dot{\rho}\|^2\} E \, \exp\{ik \int_0^t n_{1,s}(\rho(t - s)) \, ds\} V_0(\rho(t)) \, d\dot{\rho} \\
&= \int_{H_t} \exp\{\frac{ik}{2}\|\dot{\rho}\|^2\} \exp\{\frac{-k^2}{2} \int_0^t \int_0^t K(s - r, \rho(s) - \rho(r)) \, dsdr\} V_0(\rho(t)) \, d\dot{\rho}
\end{aligned}
$$

$$(6.11)$$

Now

$$
\int_0^t \int_0^t K(s - r, \rho(s) - \rho(r)) \, dsdr = \int_0^t \int_0^t \int_{\mathbf{R}^2} \exp\{i(s - r)\lambda_1 + i[\lambda_2 I_{sr}, \dot{\rho}]\} \Phi(\lambda) \, d\lambda ds dr
$$

where $\lambda = [\lambda_1 \; \lambda_2]$ and

$$
I_{sr}(t) = \begin{cases} 1 & \text{if } s < t < r \\ -1 & \text{if } r < t < s \\ 0 & \text{otherwise} \end{cases}
$$

Plugging this formula above into Equation 6.11 and expanding the exponential yields

$$E\,V_t(\rho(0)) = \int_{H_t} \exp\{\frac{ik}{2}\|\dot{\rho}\|^2\}[1 + \sum_{n=1}^{\infty}(\frac{-k^2}{2})^n/n!(\int_0^t \int_0^t \int_{\mathbf{R}^2})^j$$

$$\exp\{i(\sum_{j=1}^{n}(s_j - r_j)\lambda_{1,j} + [\lambda_{2,j}I_{s_jr_j}, \dot{\rho}])\} \prod_{j=1}^{n} \Phi(\lambda_j)\,d\lambda ds dr]V_0(\rho(t))\,d\dot{\rho}$$

$$= \int_{H_t} \exp\{\frac{ik}{2}\|\dot{\rho}\|^2\}[1 + \sum_{n=1}^{\infty}(\frac{-k^2}{2})^n/n!(\int_0^t \int_0^t \int_{\mathbf{R}^2})^j$$

$$\exp\{i\sum_{j=1}^{n}(s_j - r_j)\lambda_{1,j} - \frac{i}{2k}\|\sum_{j=1}^{n}\lambda_{2,j}I_{s_jr_j}\|^2\}$$

$$\times\ [\prod_{j=1}^{n} \Phi(\lambda_j)]V_0(\rho(t) - \frac{1}{k}\sum_{j=1}^{n}\lambda_{2,j}(s_j - r_j))\,d\lambda ds dr]\,d\dot{\rho}$$

$$= (S_tV_0)(\rho(0)) + \sum_{n=1}^{\infty}(\frac{-k^2}{2})^n/n!(\int_0^t \int_0^t \int_{\mathbf{R}^2})^n \exp\{i\sum_{j=1}^{n}(s_j - r_j)\lambda_{1,j} - \frac{i}{2k}\|\sum_{j=1}^{n}\lambda_{2,j}I_{s_jr_j}\|^2\}$$

$$\times \prod_{j=1}^{n} \Phi(\lambda_j)[S_tV_0](\rho(0) - \frac{1}{k}\sum_{j=1}^{n}\lambda_{2,j}(r_j - s_j))\,d\lambda ds dr$$

Once again the translation invariance of the finitely additive measure $d\dot{\rho}$ was used to get a closed form solution to the Feynman integral.

Up to this point, all of these calculations are exact. However it is difficult to proceed from this point without making any approximations. I assume that the propagation distance t is much greater than the outer scale L_0, hence $K(t) \approx 0$. This is a reasonable assumption in the case studied in this paper where $t = 1000\ m$ and $L_0 \approx 1\ m$. In this case the term

$$\exp\{\frac{-i}{2k}\|\sum_{j=1}^{n}\lambda_{2,j}I_{s_jr_j}\|^2\}$$

can be approximated *in the integral* as

$$\exp\{\frac{-i}{2k}\sum_{j=1}^{n}|\lambda_{2,j}|^2|s_j - r_j|\}$$

since for *most r_j and s_j*

$$[I_{s_jr_j}, I_{s_kr_k}] = 0\ , \text{for } k \neq j$$

Let $T(\rho)$ be the shift operator on $\mathcal{F}[\mathbf{R}^2]$. Then EV_t can be expressed as

$$E\,V_t\ =\ \exp\{\frac{-k^2}{2}\int_0^t \int_0^t \int_{\mathbf{R}^3} e^{i\lambda_1(r-s) - \frac{i}{2k}|\lambda_2|^2|r-s|}\Phi(\lambda)T(\frac{-1}{k}\lambda_2(r-s))\,d\lambda\,dr ds\}S_tV_0$$

$$=\ \exp\{-k^2\int_0^t \int_{\mathbf{R}^3}(t-s)e^{i\lambda_1 s - \frac{i}{2k}|\lambda_2|^2 s}\Phi(\lambda)T(\frac{-1}{k}\lambda_2 s)\,ds\,d\lambda\}S_tV_0$$

Under the Markov approximation $\Phi(\lambda) = \Phi(0, \lambda_2)$, hence this approximate mean is

$$E\,V_t = \exp\{-\frac{k^2}{2}A(0)t\}S_tV_0$$

where

$$A(0) = \int_{\mathbf{R}^2} \Phi(0, \lambda_2) \, d\lambda_2$$

which agrees with the exact result for the Markov approximation.

In the plane wave case, where $V_0 \equiv 1$ it is possible to omit the shift operator T_ρ to obtain

$$E \, V_t = \exp\{-k^2 \int_{\mathbf{R}^3} \int_0^t (t-s) e^{i\lambda_1 s - \frac{i}{2k}|\lambda_2|^2 s} \Phi(\lambda) \, d\lambda ds\}$$

This approximate solution for the mean field indicates that the mean depends on all of $\Phi(\lambda)$ and not just its integral. It also indicates that the mean will involve some phase shift.

Now let

$$\hat{K}(s, \lambda_2) = \int_{-\infty}^{\infty} e^{i\lambda_1 s} \Phi(\lambda) \, d\lambda_1$$

Then assuming that the correlation length for n_1 is much smaller than t and that the inner scale is very much larger than the wavelength of the light

$$
\begin{aligned}
E \, V_t &= \exp\{-k^2 t \int_{\mathbf{R}^2} \int_0^t \frac{t-s}{t} e^{-\frac{i}{2k}s|\lambda_2|^2} \hat{K}(s, \lambda_2) \, ds \, d\lambda_2\} \\
&\approx \exp\{-k^2 t \int_{\mathbf{R}^2} \int_0^{\infty} e^{-\frac{i}{2k}s|\lambda_2|^2} \hat{K}(s, \lambda_2) \, ds d\lambda_2\} \\
&= \exp\{\frac{-k^2}{2} t \int_{\mathbf{R}^2} \Phi(\frac{1}{2k}|\lambda_2|^2, \lambda_2) \, d\lambda_2\} \\
&\approx \exp\{\frac{-k^2}{2} t \int_{\mathbf{R}^2} \Phi(0, \lambda_2) \, d\lambda_2\} \\
&= \exp\{\frac{-k^2}{2} t A(0)\}
\end{aligned}
$$

which is consistent with the result obtained with the Markov approximation.

In this chapter the Feynman-Ito integral was used to find several approximate solutions to the forward scattering equation and an approximate expression for the mean field without the Markov approximation. The relationship between local behavior of the turbulence and the beam intensity was explored using the approximate solutions.

Bibliography

[1] S. Albeverio, Mathematical Theory of Feynman Path Integrals, Lecture Notes in Mathematics 523, Springer-Verlag, New York, 1976.

[2] S. Albeverio and R. Hoegh-Krohn, "Feynman Path Integrals and the Corresponding Method of Stationary Phase", Feynman Path Integrals, May 1978; Lecture Notes in Physics 106, Springer-Verlag, 1978, ed. S. Albeverio et. al.

[3] A. V. Balakrishnan, "Stochastic Optimization Theory in Hilbert Spaces -1", Applied Mathematics and Optimization, Vol. 1, No. 2, 1974.

[4] A. V. Balakrishnan, "Stochastic Bilinear Partial Differential Equations", Variable Structure Systems, May 1974, Lecture Notes in Economics and Mathematical Systems 111, Springer-Verlag, 1974, ed. A. Ruberti and R. R. Mohler.

[5] A. V. Balakrishnan, "On the Approximation of Ito Integrals by Band-Limited Processes", SIAM Journal on Control, Vol. 12, No. 1, May 1974.

[6] A. V. Balakrishnan, "On Abstract Stochastic Bilinear Equations", CBMS-NSF Regional Conference, 1983.

[7] A. V. Balakrishnan, "A Random Schroedinger Equation Equation: White Noise Model", Differential and Integral Equations, Vol. 1, No. 1, January 1988.

[8] A. V. Balakrishnan, Applied Functional Analysis, 2nd Edition, Springer-Verlag, New York, 1980.

[9] Yu. N. Barabanenkov, Yu. A. Kravtsov, S. M. Rytov, V. I. Tamarskii, "Status of the Theory of Propagation of Waves in a Randomly Inhomogeneous Medium", Soviet Physics Uspekhi, Vol. 13, No. 5, March-April 1971.

[10] Patrick Billingsley, Weak Convergence of Measures: Applications in Probability , Society for Industrial and Applied Mathematics, Philadelphia, 1971.

[11] Claude Chevalley, Theory of Lie Groups, Princeton University Press, Princeton, 1946.

[12] R. H. Clarke, "Analysis of Laser Beam Propagation in a Turbulent Atmosphere", AT & T Technical Journal, Vol. 64, No. 7, Sept. 1985.

[13] S. F. Clifford, "The Classical Theory of Wave Propagation in a Turbulent Medium", Laser Beam Propagation in the Atmosphere, Springer-Verlag, New York, 1978, ed. John Strohbehn.

[14] R. Dashen, "Path Integrals for Waves in Random Media", Journal of Mathematical Physics, Vol. 20, No. 5, May 1979.

[15] D. Dawson and H. Salehi, "Spatially Homgeneous Random Evolutions", Journal of Multivariate Analysis, Vol. 10, 1980.

[16] D. Dawson and G. C. Papanicolaou, "A Random Wave Process", Journal of Applied Mathematics and Optimization, Vol. 12, 1984.

[17] R. L. Fante, "Electromagnetic Beam Propagation in Turbulend Media", Proc. of the IEEE, Vol. 63, No. 12, December 1975.

[18] R. L. Fante, "Inner-Scale Size Effect on the Scintillations of Light in the Turbulent Atmosphere", Journal of the Optical Society of America, Vol. 73, No. 3, March 1983.

[19] K. Furutsu, "On the Statistical Theory of Electromagnetic Waves in a Fluctuating Medium (I)", Journal of Research of the National Bureau of Standards - D. Radio Propagation, Vol. 63D, No. 3, May-June 1963.

[20] K. Furutsu, "Statistical Theory of Wave Propagation in a Random Medium and the Irradiance Distribution Function", Journal of the Optical Society of America, Vol. 62, No. 2, Feb. 1972.

[21] A. Hald, Statistical Tables and Formulas, John Wiley and Sons, London, 1952.

[22] L. Hazareesingh and D. Kannan, "Stochastic Product Integration and Stochastic Equations", Stochastic Partial Differential Equations, Lecture Notes in Mathematics 1236, Springer-Verlag, Berlin, 1987, ed. G. Da Prato and L. Tubaro.

[23] T. Hida, Brownian Motion, Springer-Verlag, New York,1980.

[24] A. Ishimaru, The Beam Wave Case and Remote Sensing, Laser Beam Propagation in the Atmosphere, Springer-Verlag, New York, 1978, ed. J. Strohbehn.

[25] K. Ito, "Multiple Wiener Integral",Journal of the Mathematical Society of Japan, Vol. 3, No. 1, May, 1951.

[26] K. Ito, "Generalized Uniform Complex Measures in the Hilbertian Metric Space With Their Application to the Feynman Path Integral", Proceedings of the Fifth Berkeley Symposium on Mathematical Statistics and Probability, University of California Press, 1967.

[27] G. Kallianpur and R. L. Karandikar, "A Finitely Additive White Noise Approach to Nonlinear Filtering", Journal of Applied Mathematics and Optimization, Vol. 10, 1983.

[28] G. Kallianpur, K. Kannan, R. L. Karandikar, "Analytic and Sequential Feynman Integrals on Abstract Wiener and Hilbert Spaces and a Cameron-Martin Formula", Annals of the Institute Henri Poincare, Vol. 21, No. 4, 1985.

[29] G. Kallianpur, Stochastic Filtering Theory, Springer-Verlag, New York, 1980.

[30] A. Kolmogorov, In Turbulence, Classic Papers on Statistical Theory, Interscience, New York, 1961, ed. S. K. Friedlander and L. Topper.

[31] V. I. Klyatskin, "Applicability of the Approximation of a Markov Random Process in Problems Relating to the Propagation of Light in a Medium With Random Inhomogeneities", Soviet Physics Journal of Experimental and Theoretical Physics, Vol. 30, No. 3, March, 1970.

[32] V. I. Klyatskin and V. I. Tatarskii, "The Parabolic Equation Approximation for Propagation of Waves in a Medium with Random Inhomogeneities", Soviet Physics Journal of Experimental and Theoretical Physics, Vol. 31, No. 2, August, 1970.

[33] V. I. Klyatskin and V. I. Tatarskii, "A New Method of Successive Approximations in the Problem of the Propagation of Waves in a Medium Having Random Large-Scale Inhomogeneities", Radio Physics and Quantum Electronics, Vol. 14, Part 2, 1971.

[34] D. Knuth, Semi-Numerical Algorithms, The Art of Computer Programming Vol. 2, Addison Wesley, Menlo Park, CA, 1969.

[35] M. B. Marcus, "Continuity and the Central Limit Theorem for Random Trigonometric Series", Zeitschrift fur Wahrscheinlichkeitstheorie und Verwandte Gebiete, Vol. 42, 1978.

[36] A. R. Mitchell and G. Fairweather, "Improved Forms of the Alternating Direction Methods of Douglas, Peaceman and Rachford for Solving Parabolic and Elliptic Equations", Numerical Mathematics, Vol. 6, 1964.

[37] A. R. Mitchell and D. F. Griffiths, The Finite Difference Method in Partial Differential Equations, Wiley, New York, 1980.

[38] Y. Miyahara, "Infinite Diminsional Langevin Equation and Fokker-Planck Equation", Nagoya Mathematical Journal, Vol. 81, 1981.

[39] A. Mukharjea and K. Pothoven, Real and Functional Analysis, Plenum Press, New York, 1984.

[40] E. A. Novikov, "Functionals and the Random-Force Method in Turbulence Theory", Soviet Physics Journal of Experimental and Theoretical Physics, Vol. 20, No. 5, May, 1965.

[41] A. M. Obukhov, "On the Influence of Weak Atmospheric Inhomogeneities on the Propagation of Sound and Light", Izvestnya Akademia Nauka SSSR, Ser. Geofizika, No. 2, 1953.

[42] E. Platen, "An Approximation Method for a Class of Ito Processes", Lithuanian Mathematical Journal, 1981.

[43] W. H. Press, B. P. Flannery, S. A. Teukolsky, W. T. Vetterling, Numerical Recipes, Cambridge University Press, New York, 1987.

[44] A. M. Prokhorov, F. V. Bunkin, K. S. Gochelashviti, V. I. Shishov, "Propagation of Laser Radiation in Randomly Inhomogeneous Media", Soviet Physics Uspekhi, Vol. 17, No. 6, May-June, 1975.

[45] M. Reed, B. Simon, Methods of Mathematical Physics II, Academic Press, New York, 1975.

[46] S. M. Rytov, "Diffraction of Light by Ultrasonic Waves", Izvestnya Akademia Nauka SSSR, Ser. Fiz., No.2, 1937.

[47] John Strohbehn, "Line-of-Sight Wave Propagation Through the Turbulent Atmosphere", Proceedings of the IEEE, Vol. 56, No. 8, August, 1968.

[48] John Strohbehn, "On the Probability Distribution of Line-of-Sight Fluctuations of Optical Signals", Radio Science, Vol. 10, No. 1, January, 1975.

[49] John Strohbehn, ed., Laser Beam Propagation in the Atmosphere, Springer-Verlag, New York, 1978.

[50] V. I. Tatarskii, "Depolarization of Light by Turbulent Atmospheric Inhomogeneities", Radio Physics and Quantum Electronics, Vol. 10, No. 12, 1967.

[51] V. I. Tatarskii, "Laser Propagation in a Medium with Random Refractive Index Inhomogeneities in the Markov Random Process Approximation", Soviet Physics Journal of Experimental and Theoritical Physics, Vol. 29, No. 6, December, 1969.

[52] V. I. Tatarskii, Wave Propagation in a Turbulent Medium, translated by R.A. Silverman, McGraw-Hill, New York, 1961.

[53] V. I. Tatarskii, The Effects of the Turbulent Atmosphere on Wave Propagation, translated by the Israel Program for Scientific Translations, U.S. Department of Commerce, National Technical Information Service, Springfield, VA, 1971.

[54] S. Watanabe, Stochastic Differential Equations and Malliavin Calculus, Tata Institute Lectures, Springer-Verlag, Bombay, 1984.

[55] M. I. Yadrenko, Spectral Theory of Random Fields, Optimization Software, New York, 1983.

[56] K. Yosida, Functional Analysis, 6th ed., Springer-Verlag, New York, 1980.

[57] V. U. Zavorotnyi, V. I. Klyatskin, V. I. Tatarskii, "Strong Fluctuations of the Intensity of Electromagnetic Waves in Randomly Inhomogeneous Media", Soviet Physics Journal of Experimental and Theoretical Physics, Vol. 46, No. 2, August, 1977.

Appendix A

Simulation Software

In this appendix I include a description of the simulation software and brief instructions on how to use it. The software described was used to simulate the propagation of a laser beam through atmospheric turbulence on an FPS 164 array processor. Some special vectorizing features of this processor were used to improve execution time, which was usually measured in terms of hours or days. All programs written by me were written in Fortran 77.

A.1 Overview of the Program PROPAPP

The program PROPAPP generates a covariance matrix for the turbulence field and computes an approximate factorization of the matrix based on the Chuleski decomposition. The 'APP' in the program's name alludes to this approximation. This factorization is then used to generate a sequence of turbulence fields. The propagation of the laser beam through these turbulence fields is then computed using finite difference methods. Statistics of the beam at the final point, such as the coherence function, mean and irradiance function, are recorded and saved in permanent files. In addition, snapshots of the beam at points in between the initial and final point can be saved.

A.2 Instructions for Use

The behavior of the program PROPAPP is determined by parameters set in PROPAPP. A description of these parameters and their meanings is given in this section.

The parameter NEDGE defines the number of points between the center of the spatial domain being simulated and the boundary inclusive. Hence for a 33 by 33 array representation of the beam crossection NEDGE is 16.

The parameter NITER defines the number of time steps, or steps in the direction of propagation to use. In general, I used NITER = 200.

The parameter NFELDS defines the number of random turbulence fields to generate. In general this was NITER or NITER + 1.

The parameters DELTAX and DELTAY define the grid spacing in the x and y directions. The parameter DELTAT defines the 'time' step.

The parameter NRUNS determines how many Monte Carlo runs to make.

The parameter NTUSED defines how many time steps to use each random turbulence field.

The parameter NDSPFQ defines the frequency to display values of the beam crossection. The program does a display every NDSPFQ time steps.

The parameter NDORND determines whether random turbulence is generated or not.

The parameter NDONRM determines whether the norm of the beam crossection is calculated and compared with that for the homogeneous solution or not.

The parameter NDOPRT determines whether values of the beam crossection are printed out or not.

The parameter ALPHA defines the covariance of an Ornstein-Uhlenbeck process. In Chapter 5 this parameter was denoted as a.

The following parameters are physical, and are not limited to use in this program.

The parameter XK defines the wave number k.

The parameter CNSQRD defines the structure function constant C_n^2 for the random turbulence fields.

The parameter SMALL0 defines the inner scale l_0 of the turbulence.

The parameter CAPL0 defines the outer scale L_0 of the turbulence.

The program always produces the three files MEAN, COHERENC, and IRRADIAN which contain the average beam crossection, the average coherence function and the values of the irradiance function for each Monte Carlo run, all taken at the final point. The routine PUTEMF can be used to save intermediate values of the beam crossection.

A.3   Software Description

In this section I present the structure of the simulation program and a brief description of the subroutines. The module dependence diagrams for the programs PROPAPP, MAKELP and MATFLP are given in Figures A.1, A.2 and A.3. These diagrams show which routines call other routines.

The subroutines are described below in alphabetical order.

BESSJ0

BESSJ0 computes the zeroth order bessel function J_0. The algorithm was taken from [43].

CHLDCP

CHLDCP performs an approximate Chuleski decomposition to factor a covariance matrix. A finite number of pivot points can be specified, or pivot points can be chosen as the largest diagonal value.

CHL2DP

CHL2DP is a front end for CHLDCP that interprets a covariance function for a two dimensional random field as the covariance of a one dimensional random process.

FDINIT

FDINIT computes parameters for the finite difference algorithm.

GAMMA

GAMMA computes a partial gamma function. The algorithm was taken from [43].

GETVAR

GETVAR computes the variance of a random turbulence field. This routine is used to compute the covariance matrix.

GRAND

GRAND transforms a vector of uniformly distributed random deviates into a vector of Gaussian random deviates.

INTGRN

INTGRN computes the integrand used by STRFCN to compute the structure function by integrating the spectral density.

MAKCOV

MAKCOV computes the covariance operator for the random turbulence field.

MAKELP

MAKELP computes the matrix L used in producing random turbulence fields with the proper covariance $R = LL^*$

MAKOU

MAKOU transforms a sequence of independent random turbulence fields into an Ornstein-Uhlenbeck process.

MATFLP

MATFLP generates random turbulence fields using the matrices produced by MAKELP. This is somewhat of a front end for MATPRP.

MATPRP

MATPRP generates a one dimensional random process. It is used by MATFLP to produce a random field.

PROPAG

PROPAG uses finite differences to propagate the beam one step.

PROPAPP

PROPAPP is the main program described in the previous section.

PUTCOH

PUTCOH saves part of a coherence function in a permanent file.

PUTEMF

PUTEMF saves one beam crossection in a permanent file.

PUTIRR

PUTIRR saves values of the irradiance function in a permanent file.

QRTSMP

QRTSMP produces the sampling pattern, or rows to pivot on, for the approximate Chuleski decomposition used to factor the covariance matrix.

RNDVVM

RNDVVM is a system supplied routine on the FPS array processor which uses parallel processing to compute the inner products of a large number of vectors in a short amount of time.

SPECDN

SPECDN computes the value of the modified Von Karman spectral density for a given frequency.

STAEMF

STAEMF computes the average mean and coherence statistics.

STRFCN

STRFCN computes the structure function for the turbulence fields by integrating the spectral density.

VGRAND

VGRAND produces a vector of independent Gaussian random deviates with zero mean and unit variance. VGRAND is designed for use with a vectorizing processor.

VUNRND

VUNRND produces a vector of independent random deviates uniformly distributed between zero and one using a linear congruential random number generator.

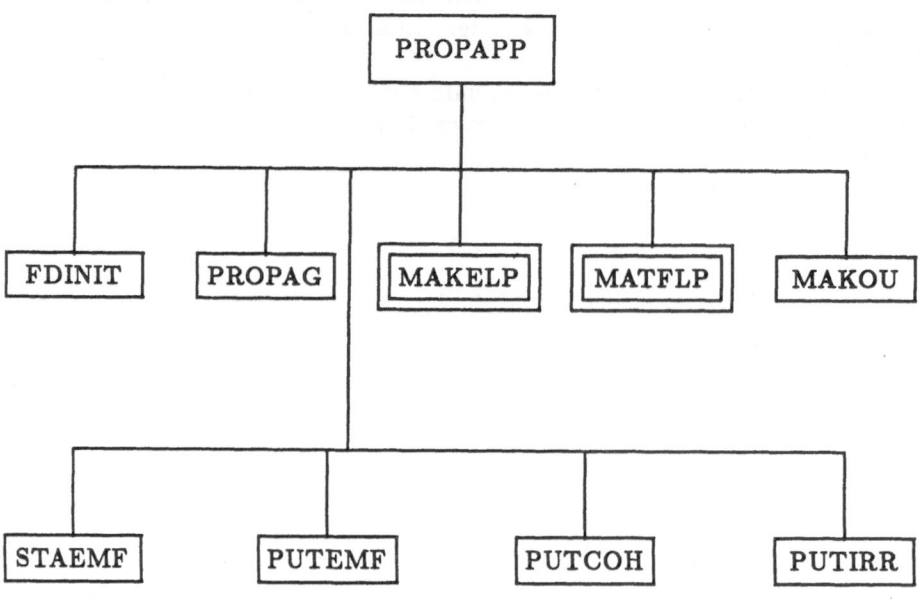

Figure A.1: Module Dependency Diagram for PROPAPP

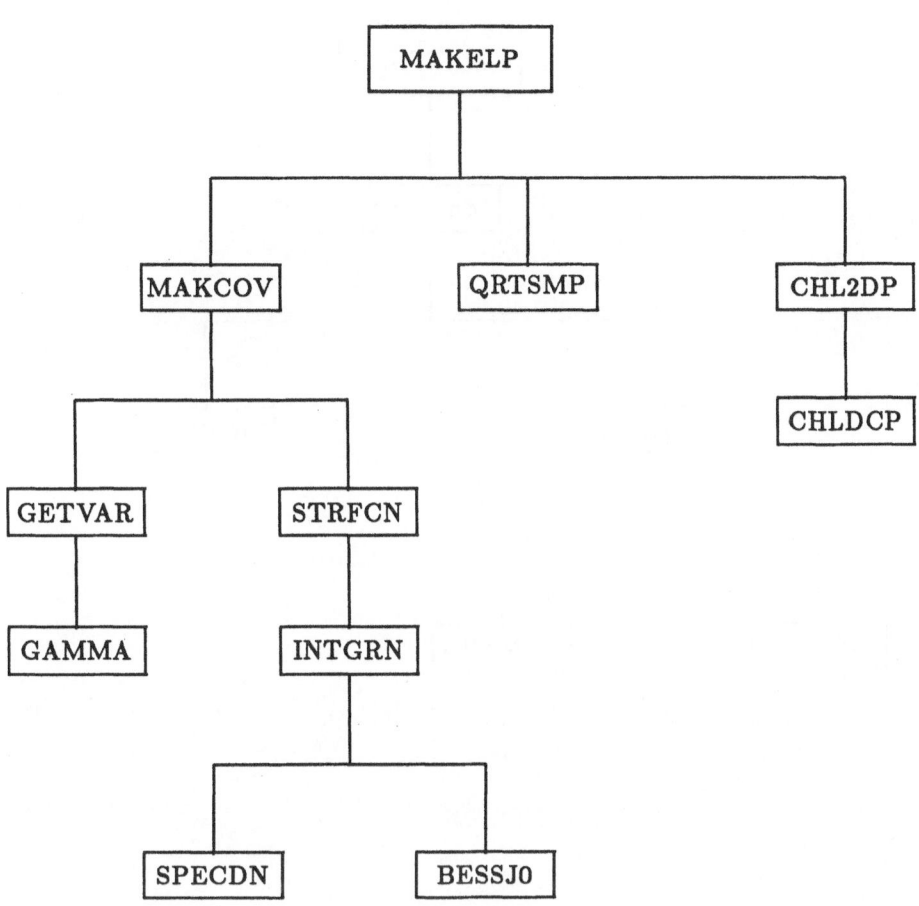

Figure A.2: Module Dependency Diagram for MAKELP

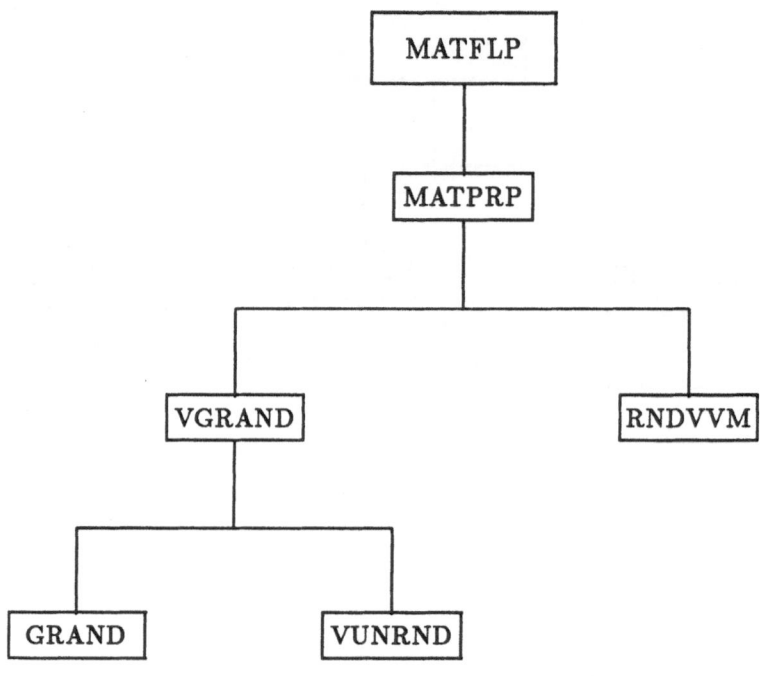

Figure A.3: Module Dependency Diagram for MATFLP

Appendix B

Additional Simulation Results

B.1 Fits of the Gamma Distribution for the Irradiance

Figures B.1 - B.6 show fits of a gamma distribution to a simulated irradiance distribution based on 8000 sample points. The fits are pretty good near the boresight of the beam, and degenerate as one moves away from it. Fits of various distributions of the irradiance are discussed in Chapter 5.

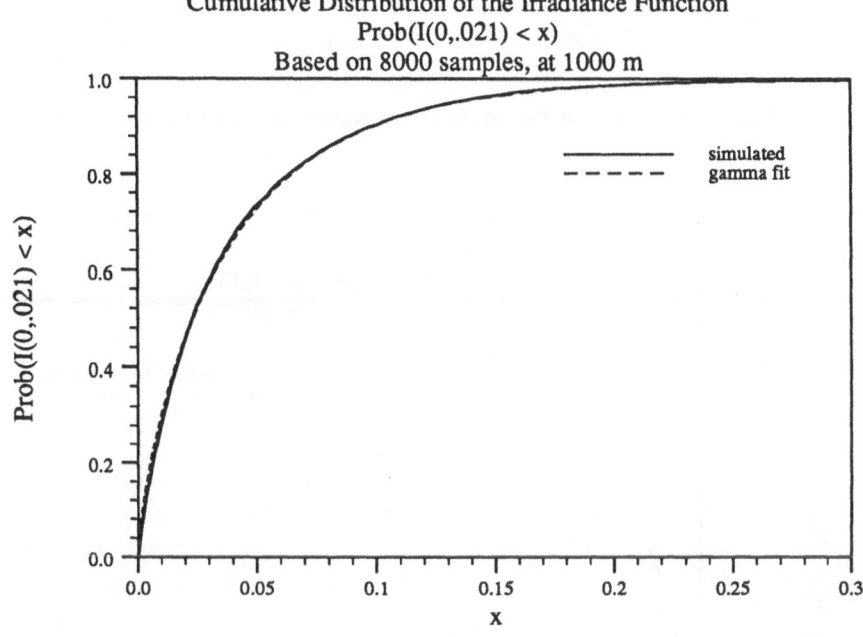

Figure B.1: Gamma Fit of Simulated Irradiance Distribution for $I(0,.021)$

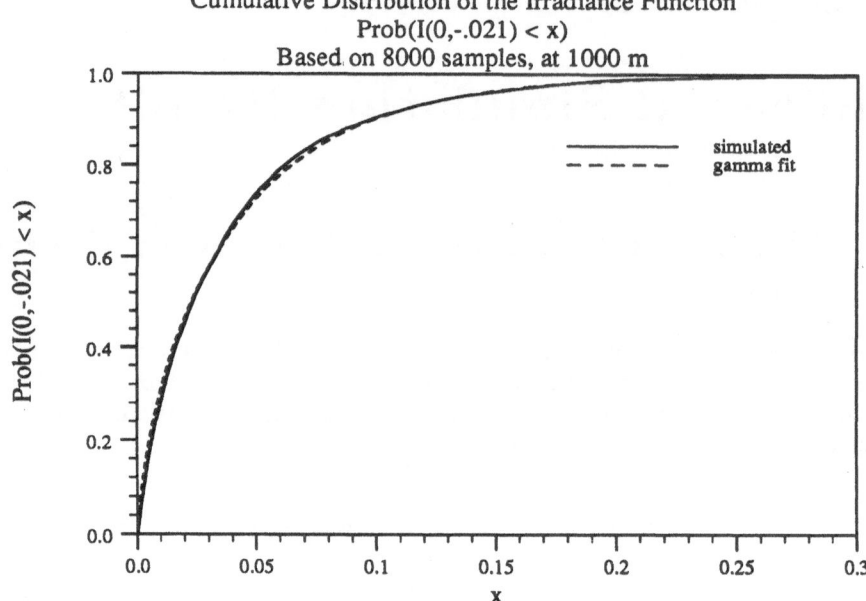

Figure B.2: Gamma Fit of Simulated Irradiance Distribution for $I(0, -.021)$

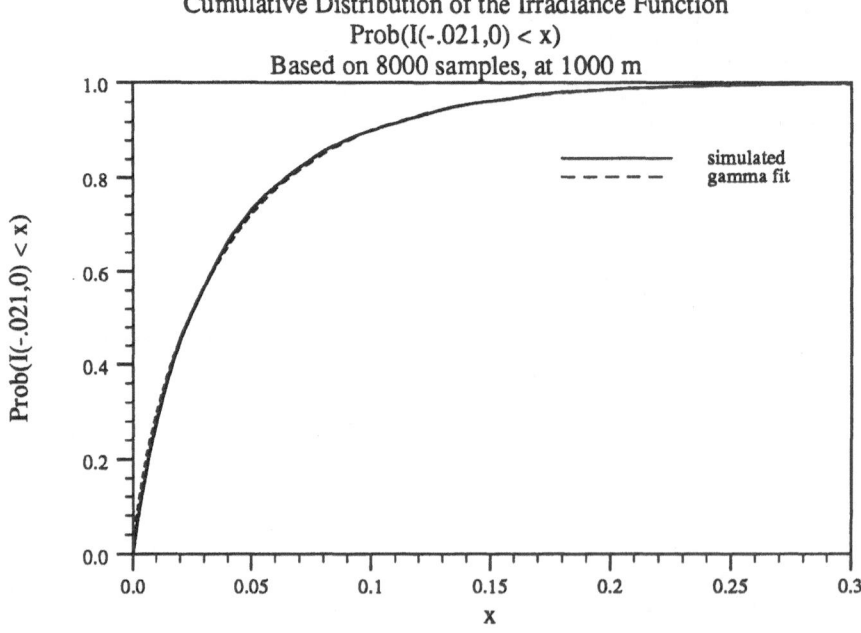

Figure B.3: Gamma Fit of Simulated Irradiance Distribution for $I(-.021, 0)$

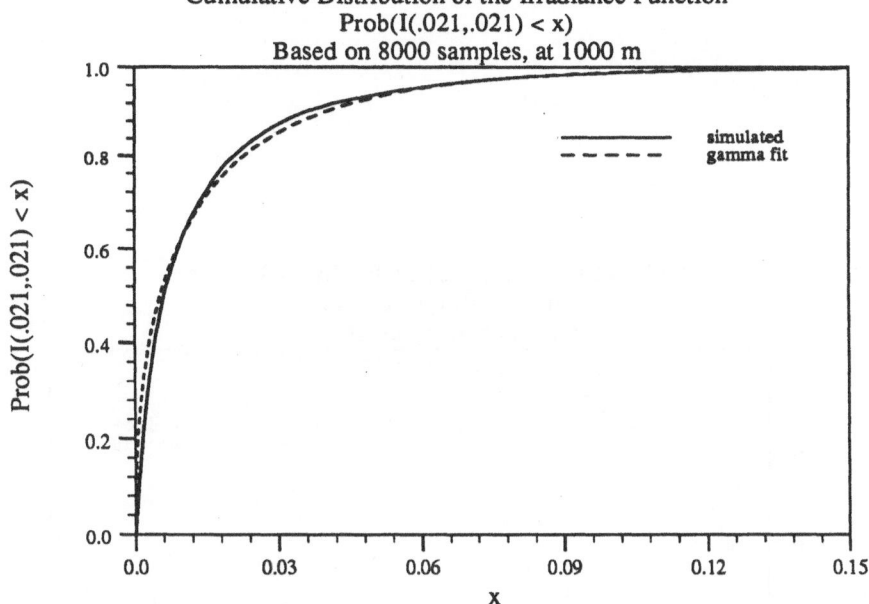

Figure B.4: Gamma Fit of Simulated Irradiance Distribution for $I(.021, .021)$

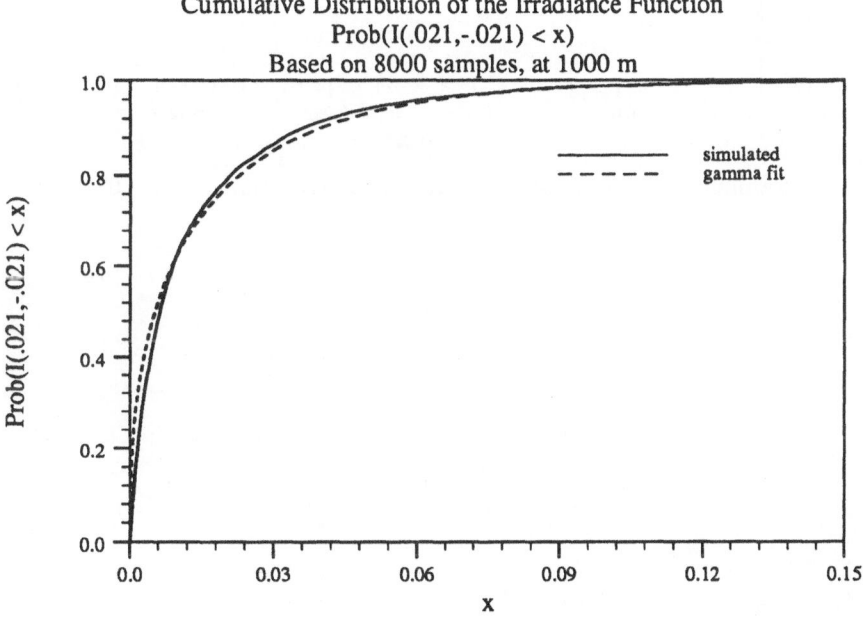

Figure B.5: Gamma Fit of Simulated Irradiance Distribution for $I(.021, -.021)$

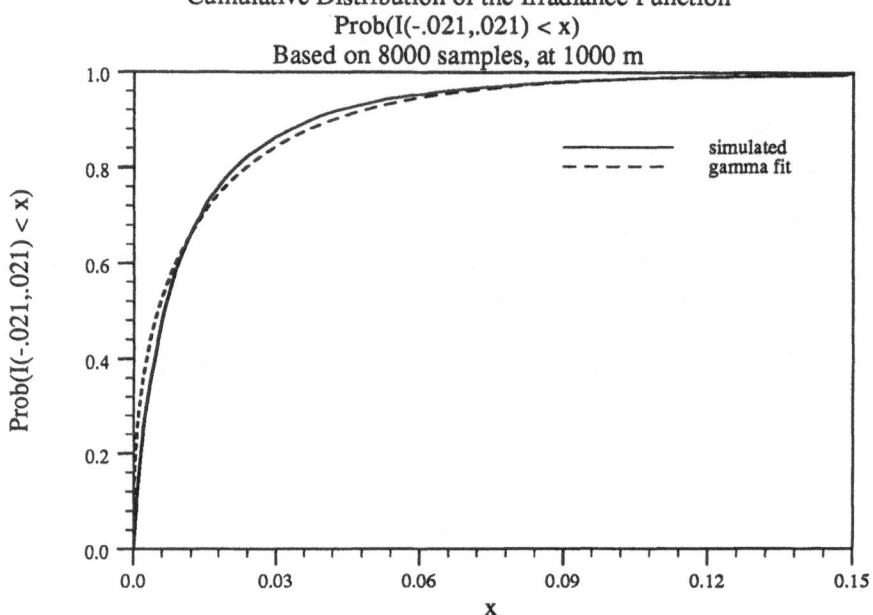

Figure B.6: Gamma Fit of Simulated Irradiance Distribution for $I(-.021, .021)$

B.2 Sample Plots Of Distorted Beams: Run 2 and Run 4

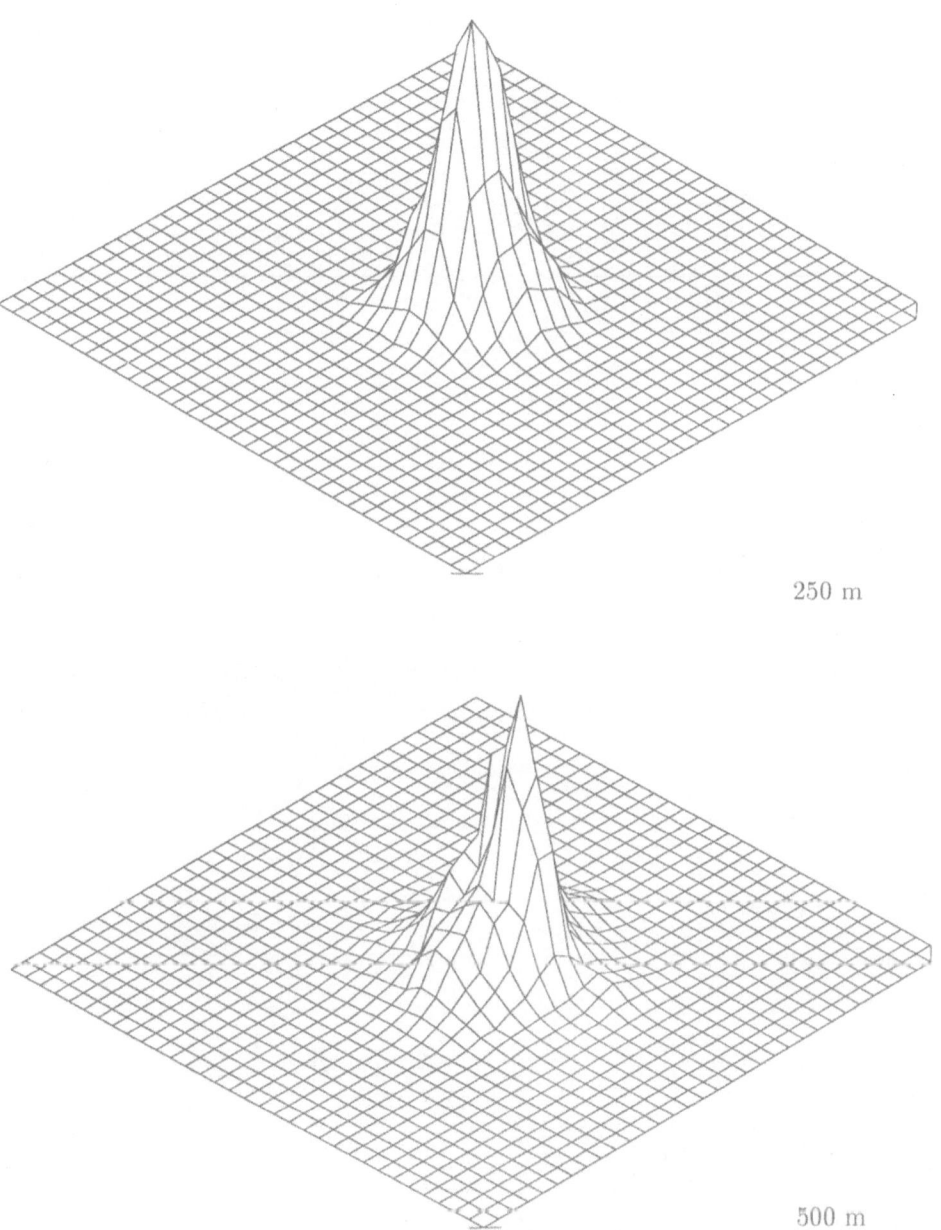

250 m

500 m

Figure B.7: Distorted Irradiance At 250 m And 500 m, 3-D Graph, Run 2

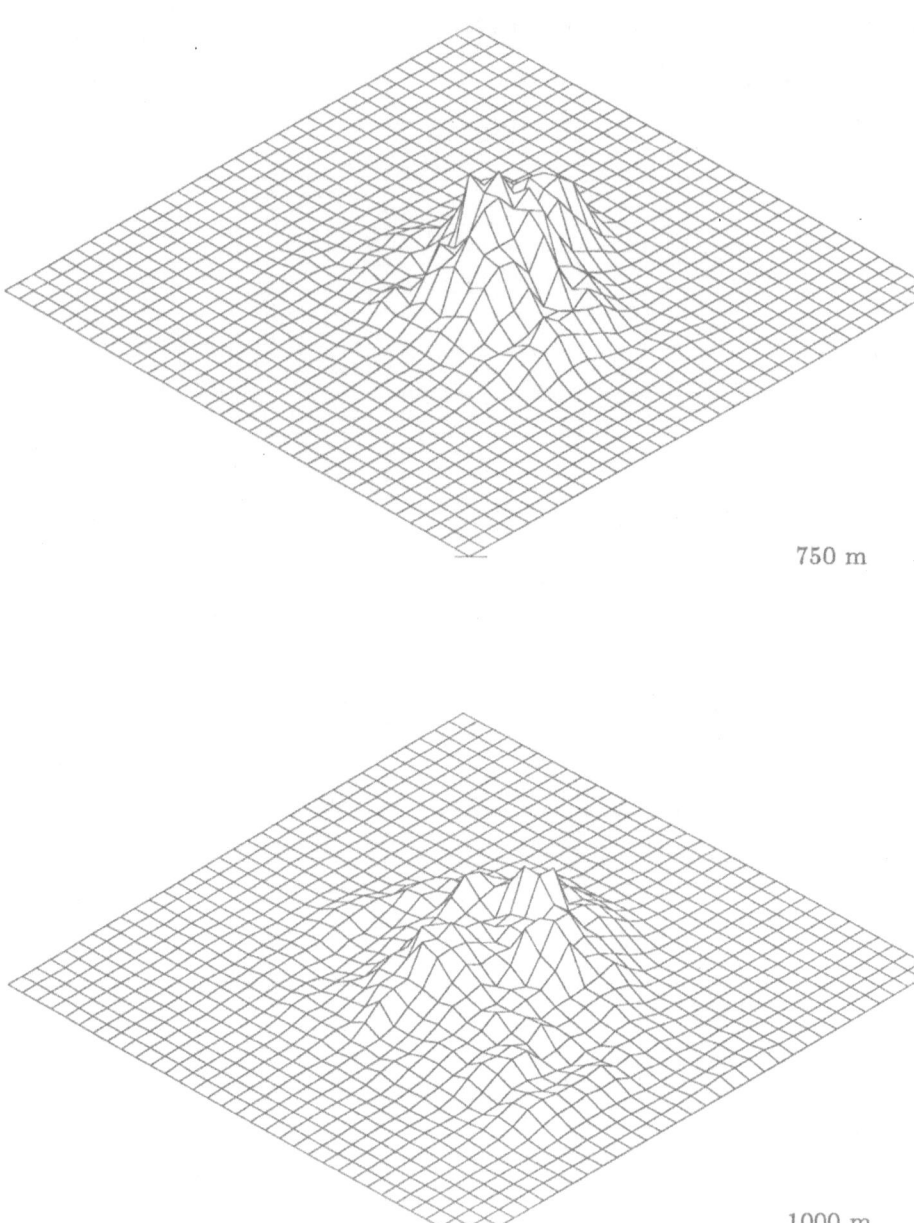

750 m

1000 m

Figure B.8: Distorted Irradiance At 750 m And 1000 m, 3-D Graph, Run 2

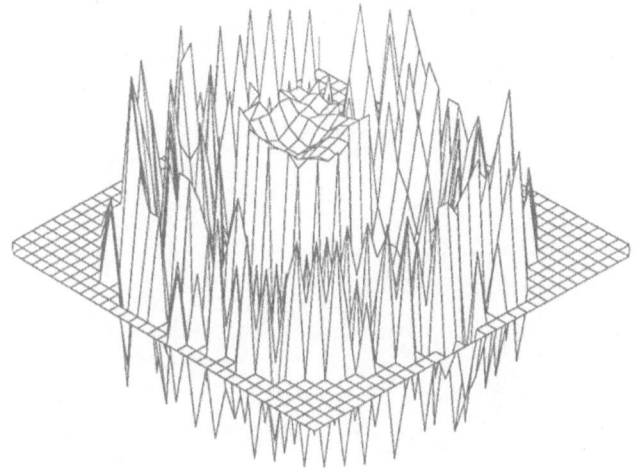

250 m

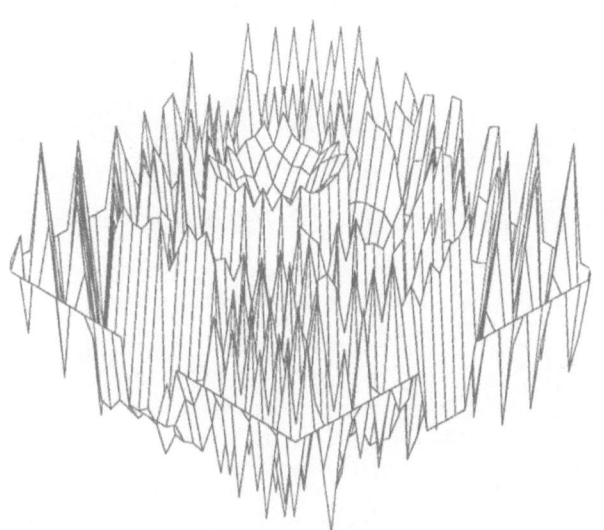

500 m

Figure B.9: Distorted Phase At 250 m And 500 m, 3-D Graph, Run 2

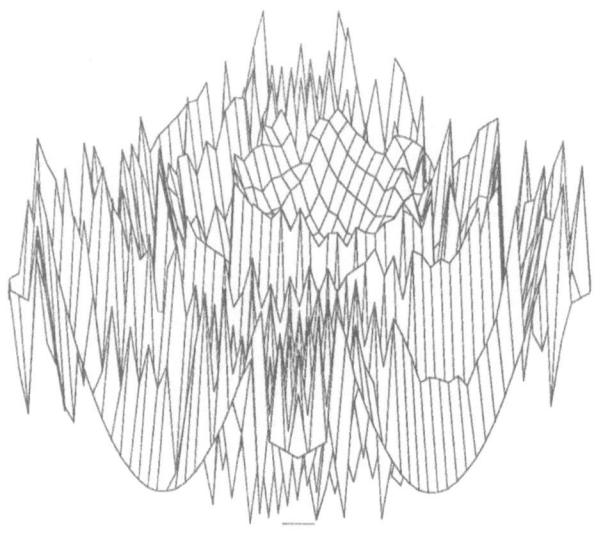

750 m

1000 m

Figure B.10: Distorted Phase At 750 m And 1000 m, 3-D Graph, Run 2

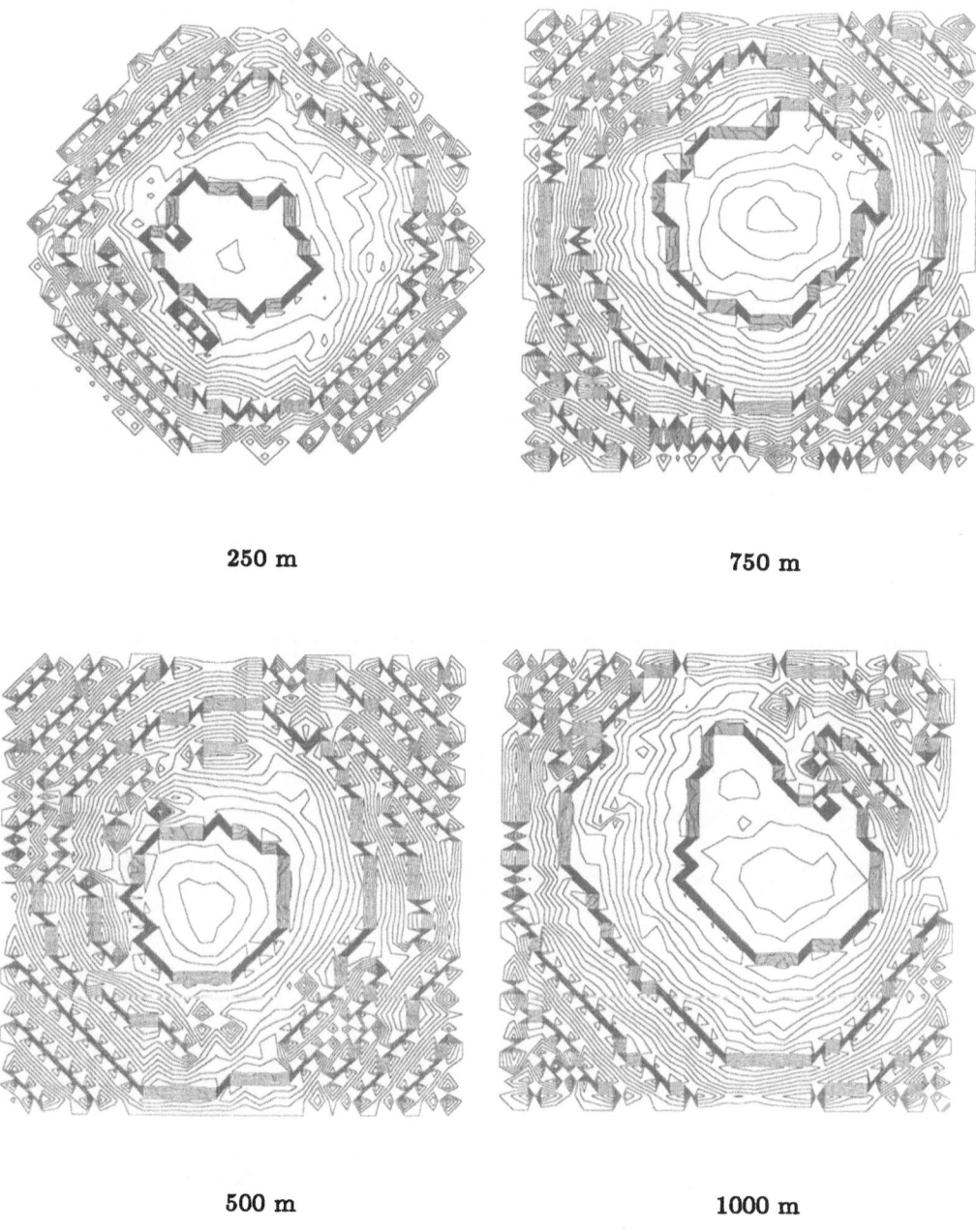

250 m

750 m

500 m

1000 m

Figure B.11: Distorted Phase, Contour Plot, step $= 2\pi/10$, Run 2

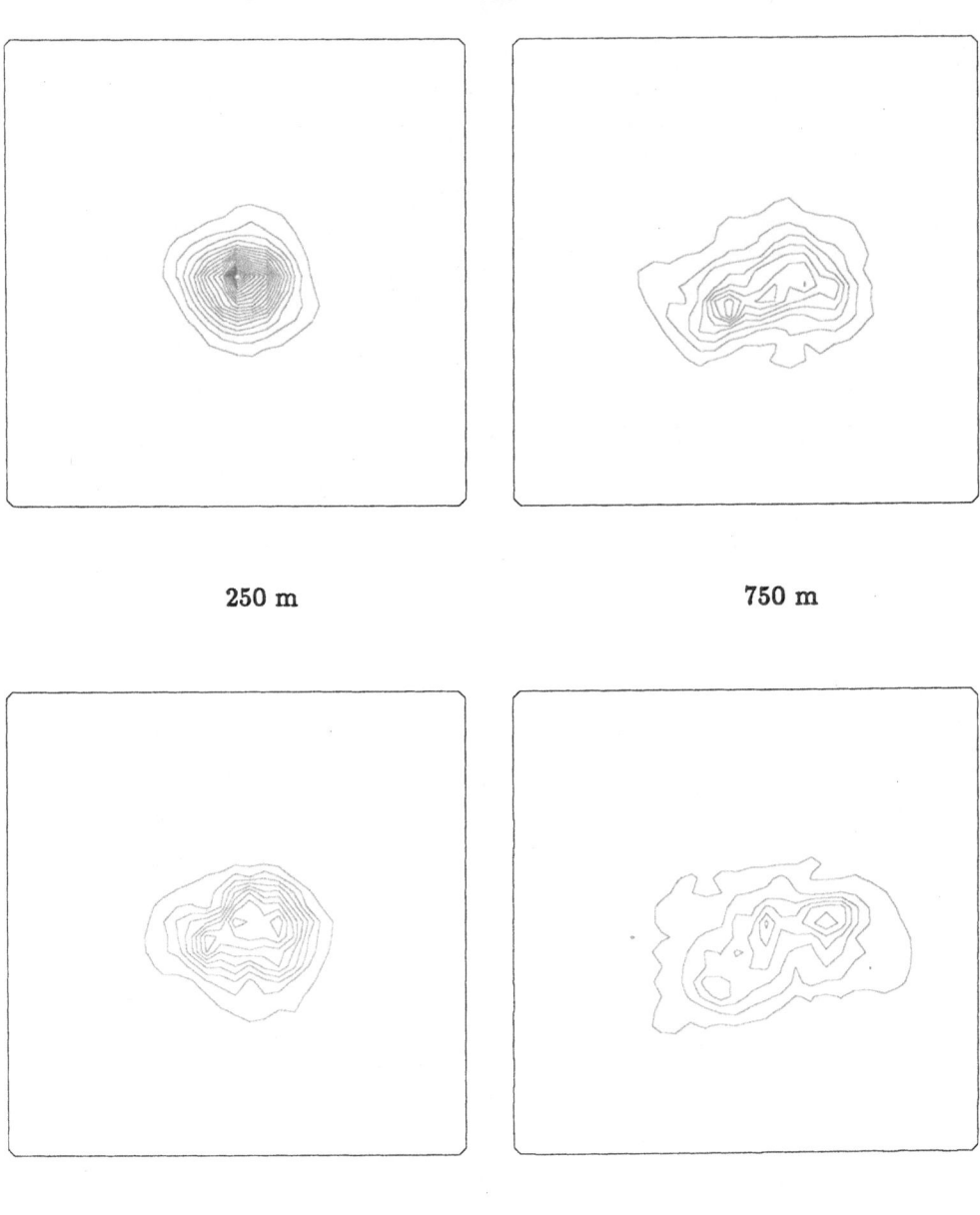

250 m

750 m

500 m

1000 m

Figure B.12: Distorted Irradiance, Contour Plot, step = .05, Run 4

137

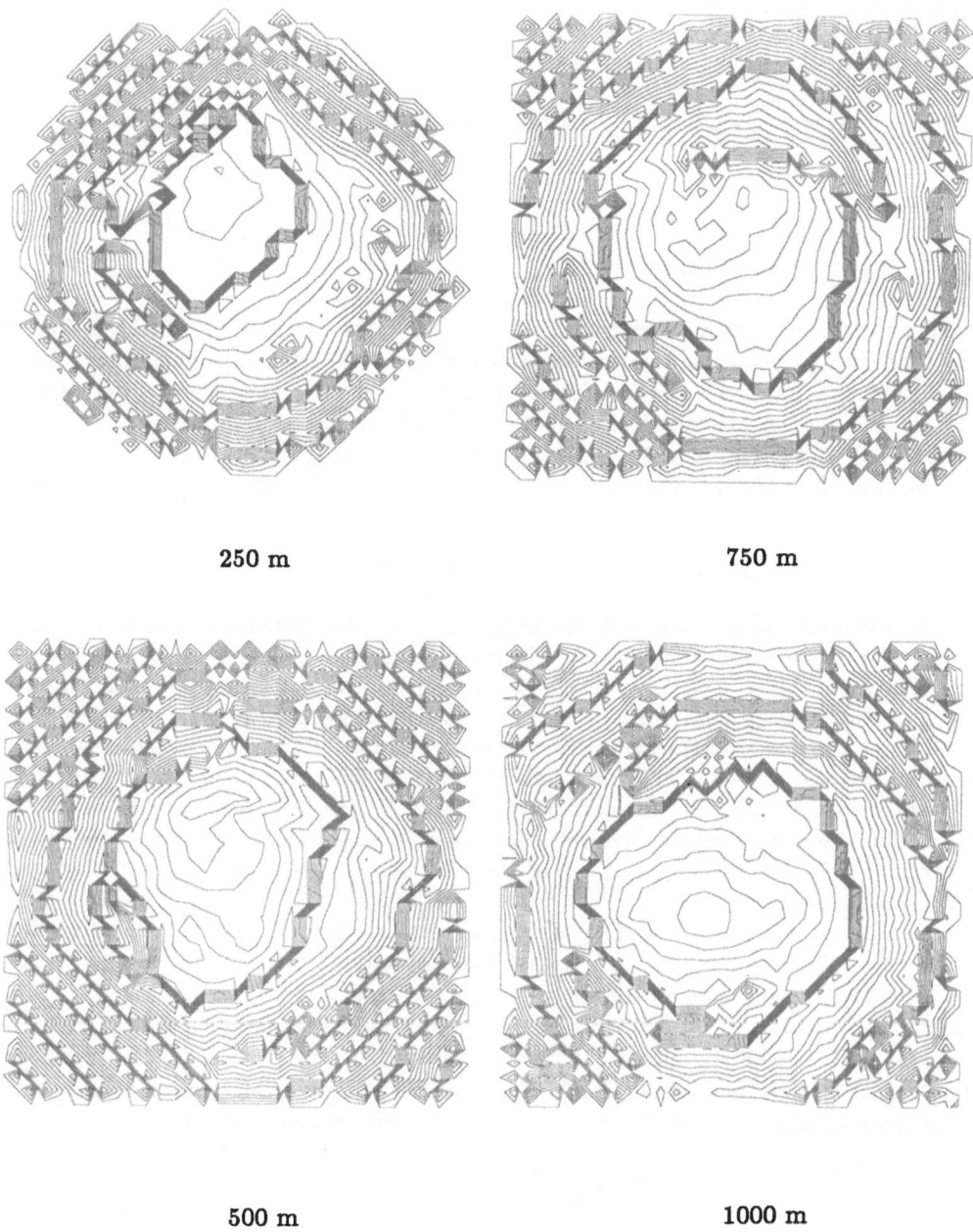

250 m

750 m

500 m

1000 m

Figure B.13: Distorted Phase, Contour Plot, step $= 2\pi/10$, Run 4

Appendix C

Simulation Verification

C.1 Introduction

In this chapter the methods and results of the simulation are tested. The statistics of the random number sequence produced are considered first. Then the accuracy of the finite difference equations is checked for the undistorted beam using several different grid spacings. Finally the product form approximation is tested by comparing the results obtained with those for finer grid spacings and by comparison with known solutions when the turbulence field is linear.

C.2 Random Number Generator

The randomness of values generated on a computer can only be considered as a statistical property of a large number of deviates. Hence it makes no sense to talk of producing 'one random number'. The tests below are performed on collections of about 10000 samples. When we say that the deviates are uncorrelated, or Gaussian, this refers to the collective properties of the samples generated, and not to the method by which they were obtained. An excellent discussion of this point is given in Knuth [34].

In order to test the random number generator, a consecutive sequence of 52928 independent Gaussian random deviates was produced and divided into five consecutive sequences of 10582 deviates each. Each of these sequences was tested for Gaussianness, correlation and mean.

The sequence was tested for Gaussianness using a chi-squared test. The real line was divided into 40 intervals, each with equal probability measure under a Normal distribution. The points to break the interval were taken from Hald [21]. Let x_j be the number of deviates that fall into interval j and let p_j be the probability measure of interval j. Let n be the number of deviates tested. From Knuth [34] the following statistic yields a random variable which asymptotically obeys a chi-squared distribution with 39 degrees of freedom if the deviates have the proper distribution.

$$V = \sum_{j=1}^{40} \frac{(x_j - np_j)^2}{np_j}$$

In our case $n = 10582$ and $p_j = .025$.

Prob	.01	.05	.1	.9	.95	.99
x	21.4	25.7	28.2	50.7	54.6	62.4

Table C.1: Chi-Squared Distribution with 39 Degrees of Freedom

Run	run 1	run 2	run 3	run 4	run 5
V	39.3	43.8	40.3	49.4	58.2

Table C.2: Chi-Squared Test Values

Values for the cumulative distribution of a chi-squared random variable with 39 degrees of freedom are given in Table C.1 and are taken from Hald [21].

Following Knuth, anything between the .1 and .9 breakpoints is accepted. Anything between the .05 and .1 or .9 breakpoints and .95 is mildly suspect, anything between the .01 and .05 or .95 and .99 breakpoints is suspect, and anything outside the .01 and .99 breakpoints is rejected.

The chi-square test values obtained in testing the random number generator are given in Table C.2. Four out of five of them are in the acceptance region and the fifth is suspect. This represents acceptable variation hence the generator passes the chi-squared test for Gaussianness.

The sample means of the random deviates should be approximately zero. The sample mean for each run itself has zero mean and a standard deviation of .0098. The sample means are given in Table C.3. The sample means are within one standard deviation of zero, hence they are acceptably small.

Finally, the correlation of the random deviates was considered. The random deviates should have a variance of 1.0 and be uncorrelated. The sample correlations for the random deviates are given in Table C.4. The standard deviation of the second moment is theoretically .0196. The standard deviation of the correlations for lags of at least one is .0098. All of the values are within one or two standard deviations, hence the sample correlations give results that are acceptably close to their expectations.

Since the random number generator passes these tests, the random turbulence fields that are generated should have the correct distributions and covariances.

Run	run 1	run 2	run 3	run 4	run 5
mean	.00328	.00328	.00712	.00471	-.00379

Table C.3: Sample Means

lag Run:	run 1	run 2	run 3	run 4	run 5
0	1.0170	1.0034	.9988	1.0195	1.0371
1	-.0029	.0094	-.0170	.0057	-.0063
2	-.0050	-.0020	-.0140	-.0159	-.0110
3	-.0154	-.0045	-.0179	.0036	.0104
4	-.0040	.0000	-.0071	.0018	.0006
5	-.0150	.0049	.0073	.0098	.0178
6	.0003	-.0136	-.0095	-.0004	.0051
7	.0064	-.0324	.0085	-.0301	.0108
8	-.0043	-.0086	.0063	-.0025	-.0109
9	-.0065	.0085	-.0031	.0046	.0083
10	-.0090	-.0002	.0001	.0026	-.0203
11	-.0119	.0053	.0200	.0068	-.0054
12	.0103	-.0024	-.0092	-.0023	.0018
13	.0017	-.0012	-.0129	-.0078	.0020
14	-.0001	-.0079	-.0081	-.0169	.0024
15	.0052	-.0046	.0023	-.0054	-.0070
16	.0113	.0043	-.0074	-.0096	.0053
17	-.0067	-.0156	.0130	-.0012	-.0125
18	-.0118	-.0021	-.0067	-.0123	.0167
19	-.0039	-.0184	-.0007	.0155	-.0149
20	.0087	.0023	-.0081	.0040	-.0141

Table C.4: Sample Correlations

C.3   Finite Difference Equations

Next the accuracy of the finite difference equations used is verified. Simulated solutions to the forward scattering equation with no distortion are compared to the calculated values. This is done by considering the L_2 norm of the error, defined by

$$\|V_{sim} - V_{act}\| = (\sum_{i,j} |V_{sim}(i,j) - V_{act}(i,j)|^2)^{\frac{1}{2}}$$

where V_{sim} and V_{act} are the simulated and actual solutions at 1000 m.

Recall that finite differences methods were used for a 31 by 31 array of points at each time step. The initial condition is taken to be

$$V_0(\rho) = e^{-|\rho|^2/10^4}$$

At the initial point, the peak magnitude is 1.0. At 1000 m the peak magnitude is about .44.

In this monograph the 'time' step was 5 m and the grid spacing was .003 m. For this sampling pattern, the norms of V_{sim} and V_{act} were 4.175 amd 4.177 respectively. The norm of the error was .01108. This represents an RMS error of .0003574 , which is highly accurate.

Next, the time step was reduced to 2.5 m and the grid spacing was kept at .003 m. In this case the error norm was .01128, with an RMS error of .0003639, which is still very accurate but a little worse than the 5 m case. Since it is cheaper to use a 5 m time step, that is what I did.

Finally, I considered the solution with a time step of 5 m and grid spacing of .005 m. For this sampling pattern, the norms of V_{sim} and V_{act} were 2.506 and 2.507 respectively and the norm of the error was .05065, with an RMS error of .001634. This is still very accurate, however this is a poorer result than with .003 grid spacing.

C.4 Product Form Approximation

In this section the accuracy of the modified product form solution is tested. First a typical turbulence field is generated. It is assumed that the random turbulence field is constant over intervals of 5 m. The time step is then reduced to 2.5 m and 1 m and the corresponding solutions are compared. The norm of all of these solutions is about 4.176. Let V_x denote the simulated solution at 1000 m with time step x. Then

$$\|V_5 - V_{2.5}\| = .0216$$

which is an RMS difference of .000698. The magnitude of the maximum difference is .00192. This is a very small difference.

For a time step of 1 m we have

$$\|V_5 - V_1\| = .0283$$

which is an RMS difference of .0009128. The magnitude of the maximum difference is .002526 . This is still a very small difference, hence the product form solution should be close to the limiting solution which is obtained as the time step converges to zero.

From Chapter 6, there is an exact expression for the solution to the forward scattering equation 2.19 when the turbulence field is merely a linear trend. Hence it is possible to simulate this case and compare the results with the actual solution.

First the simulation was run with the turbulence field set equal to

$$n_{1,t}(x,y) = qx + qy$$

For the first test I took $q = .2 \times 10^{-7}$. The graphs of the simulated values and error at 1000 m are given in Figure C.1. The error values are given below.

q	error norm	RMS error	max error
.2	.0742	.00239	.0102

As be seen from Figure C.1 much of the error comes from a kink on the boundary. If the three outermost values are ignored, the error norms are reduced as seen below.

q	error norm	RMS error	max error
.2	.0434	.00161	.00569

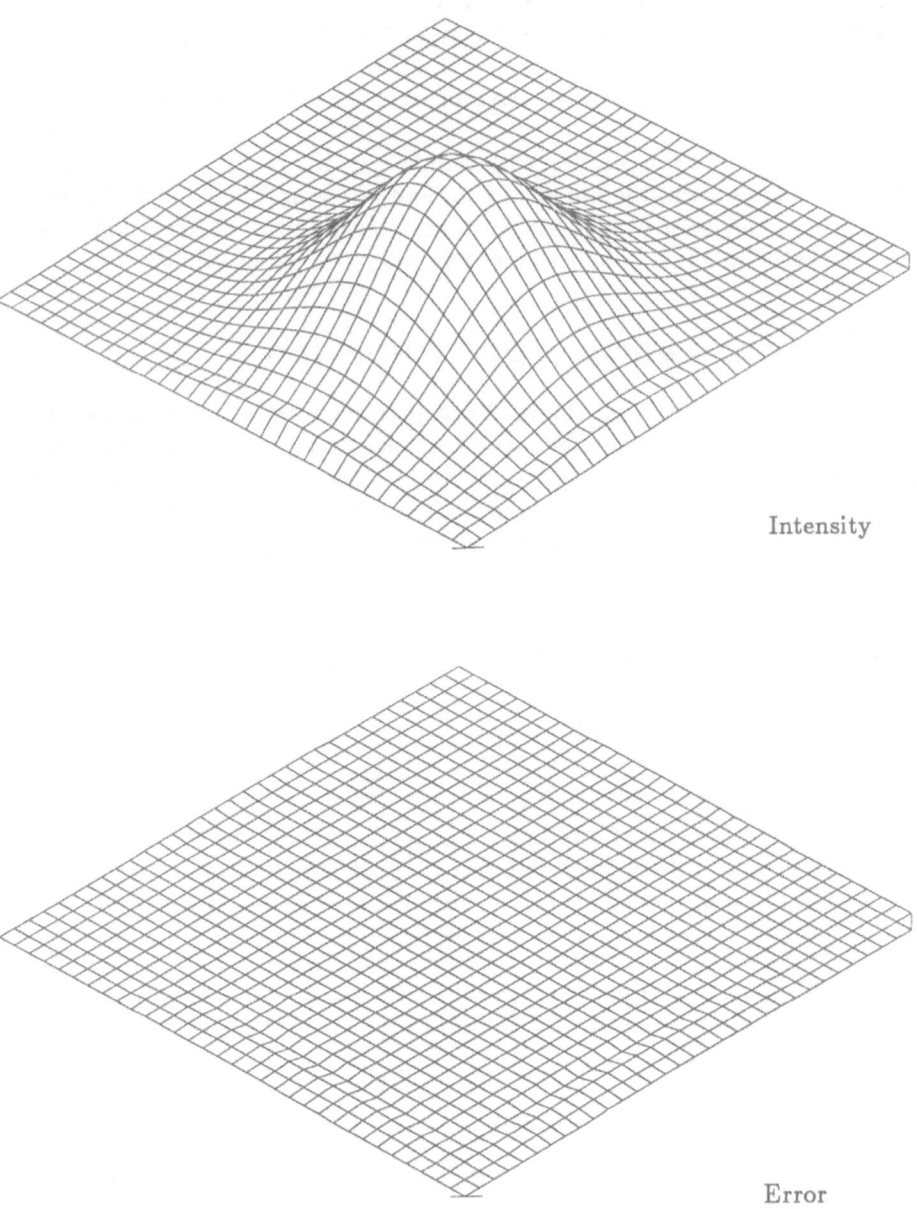

Intensity

Error

Figure C.1: Intensity and Error at 1000 m, linear input $q = .2 \times 10^{-7}$

These are sufficiently small errors for the purposes this simulation is used for, so the product formula approximation is acceptable for this input.

For the second test I took $q = .25 \times 10^{-7}$. This more intense turbulence resulted in greater error. The graphs of the simulated values and error at 1000 m are given in Figure C.2. The error values are given below.

q	error norm	RMS error	max error
.25	.114	.003687	.0166

Once again much of the error comes from a kink near the boundary. If the outermost three values are ignored, the error is reduced as seen below.

q	error norm	RMS error	max error
.25	.0709	.00263	.00927

These are slightly larger errors than I would like, but they are still acceptable. This points out the limits of the simulation, particularly with respect to inputs that cause the beam to shift off of the simulated domain. Larger values of q result in quite large errors which are not acceptable.

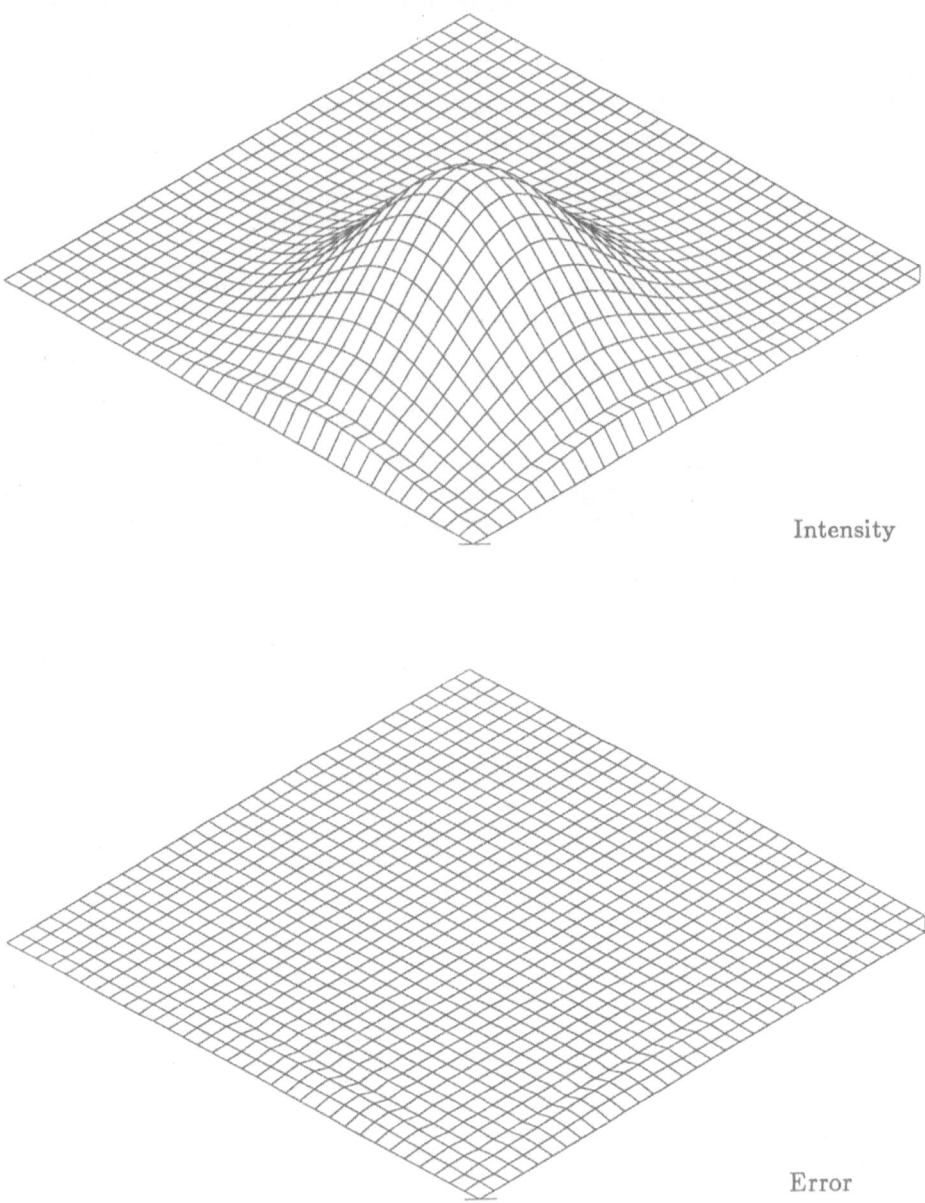

Intensity

Error

Figure C.2: Intensity and Error at 1000 m, linear input $q = .25 \times 10^{-7}$

Index

cylinder measure 14
cylinder set 14

depolarization 7

finite differences 51
forward scattering equation 8
Fresnel class 106
Furutsu-Novikov formula 10

gamma distribution 68
Greens function 8

local homogeneity 5
local isotropy 5
log-normal distribution 65

Maxwell's equations 6

nuclear operator 15

physical random variable 16
plane wave 10

random field 4
Rice-Nakagami distribution 65
Rytov's method 9

Sazanov's theorem 15
small perturbations, method of 7
smooth perturbations, method of 9
S topology 15
structure function 4,5

Von Karman 6

Lecture Notes in Control and Information Sciences

Edited by M. Thoma and A. Wyner

Vol. 81: Stochastic Optimization
Proceedings of the International Conference,
Kiew, 1984
Edited by I. Arkin, A. Shiraev, R. Wets
X, 754 pages, 1986.

Vol. 82: Analysis and Algorithms
of Optimization Problems
Edited by K. Malanowski, K. Mizukami
VIII, 240 pages, 1986.

Vol. 83: Analysis and Optimization
of Systems
Proceedings of the Seventh International
Conference of Analysis and Optimization
of Systems
Antiba, June 26-27, 1986
Edited by A. Bensoussan, J. L. Lions
XVI, 901 pages, 1986.

Vol. 84: System Modelling
and Optimization
Proceedings of the 12th IFIP Conference
Budapest, Hungary, September 2–6, 1985
Edited by A. Prékopa, J. Szelezsán, B. Strazicky
XII, 1046 pages, 1986.

Vol. 85: Stochastic Processes
in Underwater Acoustics
Edited by Charles R. Baker
V, 205 pages, 1986.

Vol. 86: Time Series and
Linear Systems
Edited by Sergio Bittanti
XVII, 243 pages, 1986.

Vol. 87: Recent Advances in
System Modelling and
Optimization
Proceedings of the IFIP-WG 7/1
Working Conference
Santiago, Chile, August 27-31, 1984
Edited by L. Contesse, R. Correa, A. Weintraub
IV, 199 pages, 1987.

Vol. 88: Bruce A. Francis
A Course in H_∞ Control Theory
XI, 156 pages, 1987.

Vol. 88: Bruce A. Francis
A Course in H_∞ Control Theory
X, 150 pages, 1987.
Corrected · 1st printing 1987

Vol. 89: G. K. H. Pang/A. G. J. McFarlane
An Expert System Approach to
Computer-Aided Design
of Multivariable Systems
XII, 223 pages, 1987.

Vol. 90: Singular Perturbations
and Asymptotic Analysis
in Control Systems
Edited by P. Kokotovic,
A. Bensoussan, G. Blankenship
VI, 419 pages, 1987.

Vol. 91 Stochastic Modelling
and Filtering
Proceedings of the IFIP-WG 7/1
Working Conference
Rome, Italy, Decembre 10-14, 1984
Edited by A. Germani
IV, 209 pages, 1987.

Vol. 92: L. T. Grujić, A. A. Martynyuk,
M. Ribbens-Pavella
Large-Scale Systems Stability Under
Structural and Singular Perturbations
XV, 366 pages, 1987.

Vol. 93: K. Malanowski
Stability of Solutions to Convex
Problems of Optimization
IX, 137 pages, 1987.

Vol. 94: H. Krishna
Computational Complexity
of Bilinear Forms
Algebraic Coding Theory and
Applications to Digital
Communication Systems
XVIII, 166 pages, 1987.

Vol. 95: Optimal Control
Proceedings of the Conference on
Optimal Control and Variational Calculus
Oberwolfach, West-Germany, June 15-21, 1986
Edited by R. Bulirsch, A. Miele, J. Stoer
and K. H. Well
XII, 321 pages, 1987.

Vol. 96: H. J. Engelbert/W. Schmidt
Stochastic Differential Systems
Proceedings of the IFIP-WG 7/1
Working Conference
Eisenach, GDR, April 6-13, 1986
XII, 381 pages, 1987.

Lecture Notes in Control and Information Sciences

Edited by M. Thoma and A. Wyner

Vol. 97: I. Lasiecka/R. Triggiani (Eds.)
Control Problems for Systems
Described by Partial Differential Equations
and Applications
Proceedings of the IFIP-WG 7.2
Working Conference
Gainesville, Florida, February 3-6, 1986
VIII, 400 pages, 1987.

Vol. 98: A. Aloneftis
Stochastic Adaptive Control
Results and Simulation
XII, 120 pages, 1987.

Vol. 99: S. P. Bhattacharyya
Robust Stabilization Against
Structured Perturbations
IX, 172 pages, 1987.

Vol. 100: J. P. Zolésio (Editor)
Boundary Control and Boundary Variations
Proceedings of the IFIP WG 7.2 Conference
Nice, France, June 10-13, 1987
IV, 398 pages, 1988.

Vol. 101: P. E. Crouch,
A. J. van der Schaft
Variational and Hamiltonian
Control Systems
IV, 121 pages, 1987.

Vol. 102: F. Kappel, K. Kunisch,
W. Schappacher (Eds.)
Distributed Parameter Systems
Proceedings of the 3rd International Conference
Vorau, Styria, July 6–12, 1986
VII, 343 pages, 1987.

Vol. 103: P. Varaiya, A. B. Kurzhanski (Eds.)
Discrete Event Systems:
Models and Applications
IIASA Conference
Sopron, Hungary, August 3-7, 1987
IX, 282 pages, 1988.

Vol. 104: J. S. Freudenberg/D. P. Looze
Frequency Domain Properties of Scalar
and Multivariable Feedback Systems
VIII, 281 pages, 1988.

Vol. 105: Ch. I. Byrnes/A. Kurzhanski (Eds.)
Modelling and Adaptive Control
Proceedings of the IIASA Conference
Sopron, Hungary, July 1986
V, 379 pages, 1988.

Vol. 106: R. R. Mohler (Editor)
Nonlinear Time Series and
Signal Processing
V, 143 pages. 1988.

Vol. 107: Y. T. Tsay, L.-S. Shieh, St. Barnett
Structural Analysis and Design
of Multivariable Systems
An Algebraic Approach
VIII, 208 pages, 1988.

Vol. 108: K. J. Reinschke
Multivariable Control
A Graph-theoretic Approach
274 pages, 1988.

Vol. 109: M. Vukobratović/R. Stojić
Modern Aircraft Flight Control
VI, 288 pages, 1988.

Vol. 110: In preparation

Vol. 111: A. Bensoussan, J. L. Lions (Eds.)
Analysis and Optimization
of Systems
XIV, 1175 pages, 1988.

Vol. 112: Vojislav Kecman
State-Space Models of Lumped
and Distributed Systems
IX, 280 pages, 1988

Vol. 113: M. Iri, K. Yajima (Eds.)
System Modelling and Optimization
Proceedings of the 13th IFIP Conference
Tokyo, Japan, Aug. 31 – Sept. 4, 1987
IX, 787 pages, 1988.

Vol. 114: A. Bermúdez (Editor)
Control of Partial Differential Equations
Proceedings of the IFIP WG 7.2
Working Conference
Santiago de Compostela, Spain, July 6–9, 1987
IX, 318 pages, 1989

Vol. 115: H.J. Zwart
Geometric Theory for Infinite
Dimensional Systems
VIII, 156 pages, 1989.

Vol. 116: M.D. Mesarovic, Y. Takahara
Abstract Systems Theory
VIII, 439 pages, 1989

Lecture Notes in Control and Information Sciences

Edited by M. Thoma and A. Wyner

Vol. 117: K.J. Hunt
Stochastic Optimal Control Theory
with Application in Self-Tuning Control
X, 308 pages, 1989.

Vol. 118: L. Dai
Singular Control Systems
IX, 332 pages, 1989

Vol. 119: T. Başar, P. Bernhard
Differential Games and Applications
VII, 201 pages, 1989

Vol. 120: L. Trave, A. Titli, A. M. Tarras
Large Scale Systems:
Decentralization, Structure Constraints
and Fixed Modes
XIV, 384 pages, 1989

Vol. 121: A. Blaquière (Editor)
Modeling and Control of Systems
in Engineering, Quantum Mechanics,
Economics and Biosciences
Proceedings of the Bellman Continuum
Workshop 1988, June 13–14, Sophia Antipolis, France
XXVI, 519 pages, 1989

Vol. 122: J. Descusse, M. Fliess, A. Isidori,
D. Leborgne (Eds.)
New Trends in Nonlinear Control Theory
Proceedings of an International
Conference on Nonlinear Systems,
Nantes, France, June 13–17, 1988
VIII, 528 pages, 1989

Vol. 123: C. W. de Silva, A. G. J. MacFarlane
Knowledge-Based Control with
Application to Robots
X, 196 pages, 1989

Vol. 124: A. A. Bahnasawi, M. S. Mahmoud
Control of Partially-Known
Dynamical Systems
XI, 228 pages, 1989

Vol. 125: J. Simon (Ed.)
Control of Boundaries and Stabilization
Proceedings of the IFIP WG 7.2 Conference
Clermont Ferrand, France, June 20–23, 1988
IX, 266 pages, 1989

Vol. 126: N. Christopeit, K. Helmes
M. Kohlmann (Eds.)
Stochastic Differential Systems
Proceedings of the 4th Bad Honnef Conference
June 20–24, 1988
IX, 342 pages, 1989

Vol.127: C. Heij
Deterministic Identification
of Dynamical Systems
VI, 292 pages, 1989

Vol. 128: G. Einarsson, T. Ericson,
I. Ingemarsson, R. Johannesson,
K. Zigangirov, C.-E. Sundberg
Topics in Coding Theory
VII, 176 pages, 1989

Vol. 129: W. A.Porter, S. C. Kak (Eds.)
Advances in Communications and
Signal Processing,
Proceedings of an International Conference
Baton Rouge, Louisiana, October 19–21, 1988
VI, 376 pages, 1989.

Vol. 130: W. A. Porter, S. C. Kak,
J. L. Aravena (Eds.)
Advances in Computing and Control
Proceedings of an International Conference
Baton Rouge, Louisiana, October 19–21, 1988,
VI, 367 pages, 1989

Vol. 131: S. M. Joshi
Control of Large Flexible Space Structures
IX, 196 pages, 1989.

Vol. 132: W.-Y. Ng
Interactive Multi-Objective Programming
as a Framework for Computer-Aided Control
System Design
XV, 182 pages, 1989.

Vol. 133: R. P. Leland
Stochastic Models for Laser Propagation
in Atmospheric Turbulence
VII, 145 pages, 1989.